A TREATISE ON ADULTERATIONS OF FOOD

AND CULINARY POISONS

By

Fredrick Accum

tion et mise en page : Culturea (Hérault, 34)
rouvez notre catalogue sur http://culturea.fr/
ntact : infos@culturea.fr
primé en Allemagne par Books on Demand Gmbh
sign typographique et layout : Derek Murphy et Reedsy
N : 9791041985692
e de parution : février 2024

PRELIMINARY OBSERVATIONS.

Of all the frauds practised by mercenary dealers, there is none more reprehensible, and at the same time more prevalent, than the sophistication of the various articles of food.

This unprincipled and nefarious practice, increasing in degree as it has been found difficult of detection, is now applied to almost every commodity which can be classed among either the necessaries or the luxuries of life, and is carried on to a most alarming extent in every part of the United Kingdom.

It has been pursued by men, who, from the magnitude and apparent respectability of their concerns, would be the least obnoxious to public suspicion; and their successful example has called forth, from among the retail dealers, a multitude of competitors in the same iniquitous course.

To such perfection of ingenuity has this system of adulterating food arrived, that spurious articles of various kinds are every where to be found, made up so skilfully as to baffle the discrimination of the most experienced judges.

Among the number of substances used in domestic economy which are now very generally found sophisticated, may be distinguished— tea, coffee, bread, beer, wine, spiritous liquors, salad oil, pepper, vinegar, mustard, cream, and other articles of subsistence.

Indeed, it would be difficult to mention a single article of food which is not to be met with in an adulterated state; and there are some substances which are scarcely ever to be procured genuine.

Some of these spurious compounds are comparatively harmless when used as food; and as in these cases merely substances of inferior value are substituted for more costly and genuine ingredients, the sophistication, though it may affect our purse, does not injure our health. Of this kind are the manufacture of factitious pepper, the adulterations of mustard, vinegar, cream, &c. Others, however, are highly deleterious; and to this class belong the adulterations of beer, wines, spiritous liquors, pickles, salad oil, and many others.

There are particular chemists who make it a regular trade to supply drugs or nefarious preparations to the unprincipled brewer of porter or ale; others perform the same office to the wine and spirit merchant; and others again to the grocer and the oilman. The operators carry on their processes chiefly in secresy, and under some delusive firm, with the ostensible denotements of a fair and lawful establishment.

These illicit pursuits have assumed all the order and method of a regular trade; they may severally claim to be distinguished as an *art and mystery*; for the workmen employed in them are often wholly ignorant of the nature of the substances which pass through their hands, and of the purposes to which they are ultimately applied.

To elude the vigilance of the inquisitive, to defeat the scrutiny of the revenue officer, and to ensure the secresy of these mysteries, the processes are very ingeniously divided and subdivided among individual operators, and the manufacture is purposely carried on in separate establishments. The task of proportioning the ingredients for use is assigned to one individual, while the composition and preparation of them may be said to form a distinct part of the business, and is entrusted to another workman. Most of the articles are transmitted to the consumer in a disguised state, or in such a form that their real nature cannot possibly be detected by the unwary. Thus the extract of *coculus indicus*, employed by fraudulent manufacturers of malt-liquors to impart an intoxicating quality to porter or ales, is known in the market by the name of *black extract*, ostensibly destined for the use of tanners and dyers. It is obtained by boiling the berries of the coculus indicus in water, and converting, by a subsequent evaporation, this decoction into a stiff black tenacious mass, possessing, in a high degree, the narcotic and intoxicating quality of the poisonous berry from which it is prepared. Another substance, composed of extract of quassia and liquorice juice, used by fraudulent brewers to economise both malt and hops, is technically called *multum*.

The quantities of coculus indicus berries, as well as of black extract, imported into this country for adulterating malt liquors, are enormous. It forms a considerable branch of commerce in the hands of a few brokers: yet, singular as it may seem, no inquiry appears to have been hitherto made by the officers of the revenue respecting its application. Many other substances employed in the adulteration of beer, ale, and spiritous liquors, are in a similar manner intentionally disguised; and of the persons by whom they are purchased, a great number are totally unacquainted with their nature or composition.

An extract, said to be innocent, sold in casks, containing from half a cwt. to five cwt. by the brewers' druggists, under the name of *bittern*, is composed of calcined sulphate of iron (copperas), extract of coculus indicus berries, extract of quassia, and Spanish liquorice.

It would be very easy to adduce, in support of these remarks, the testimony of numerous individuals, by whom I have been professionally engaged to examine certain mixtures, said to be perfectly innocent, which are used in very extensive manufactories of the above description. Indeed, during the long period devoted to the practice of my profession, I have had abundant reason to be convinced that a vast number of dealers, of the highest respectability, have vended to their customers articles absolutely poisonous, which they themselves considered as harmless, and which they would not have offered for sale, had they been apprised of the spurious and pernicious nature of the compounds, and of the purposes to which they were destined.

For instance, I have known cases in which brandy merchants were not aware that the substance which they frequently purchase under the delusive name of *flash*, for strengthening and clarifying spiritous liquors, and which is held out as consisting of burnt sugar and isinglass only, in the form of an extract, is in reality a compound of sugar, with extract of capsicum; and that to the acrid and pungent qualities of the capsicum is to be ascribed the heightened flavour of brandy and rum, when coloured with the above-mentioned matter.

In other cases the ale-brewer has been supplied with ready-ground coriander seeds, previously mixed with a portion of *nux vomica* and quassia, to give a bitter taste and narcotic property to the beverage.

The retail venders of mustard do not appear to be aware that mustard seed alone cannot produce, when ground, a powder of so intense and brilliant a colour as that of the common mustard of commerce. Nor would the powder of real mustard, when mixed with salt and water, without the addition of a portion of pulverised capsicum, keep for so long a time as the mustard usually offered for sale.

Many other instances of unconscious deceptions might be mentioned, which were practised by persons of upright and honourable minds.

It is a painful reflection, that the division of labour which has been so instrumental in bringing the manufactures of this country to their present flourishing state, should have also tended to conceal and facilitate the fraudulent practices in question; and that from a correspondent ramification of commerce into a multitude of distinct branches, particularly in the metropolis and the large towns of the empire, the traffic in adulterated commodities should find its way through so many circuitous channels, as to defy the most scrutinising endeavour to trace it to its source.

It is not less lamentable that the extensive application of chemistry to the useful purposes of life, should have been perverted into an auxiliary to this nefarious traffic. But, happily for the science, it may, without difficulty, be converted into a means of detecting the abuse; to effect which, very little chemical skill is required; and the course to be pursued forms the object of the following pages.

The baker asserts that he does not put alum into bread; but he is well aware that, in purchasing a certain quantity of flour, he must take a sack of *sharp whites* (a term given to flour contaminated with a quantity of alum), without which it would be impossible for him to produce light, white, and porous bread, from a half-spoiled material.

The wholesale mealman frequently purchases this spurious commodity, (which forms a separate branch of business in the hands of certain individuals,) in order to enable himself to sell his decayed and half-spoiled flour.

Other individuals furnish the baker with alum mixed up with salt, under the obscure denomination of *stuff*. There are wholesale manufacturing chemists, whose sole business is to crystallise alum, in such a form as will adapt this salt to the purpose of being mixed in a crystalline state with the crystals of common salt, to disguise the character of the compound. The mixture called*stuff*, is composed of one part of alum, in minute crystals, and three of common salt. In many other trades a similar mode of proceeding prevails. Potatoes are soaked in water to augment their weight.

The practice of sophisticating the necessaries of life, being reduced to systematic regularity, is ranked by public opinion among other mercantile pursuits; and is not only regarded with less disgust than formerly, but is almost generally esteemed as a justifiable way to wealth.

It is really astonishing that the penal law is not more effectually enforced against practices so inimical to the public welfare. The man who robs a fellow subject of a few shillings on the high-way, is sentenced to death; while he who distributes a slow poison to a whole community, escapes unpunished.

It has been urged by some, that, under so vast a system of finance as that of Great Britain, it is expedient that the revenue should be collected in large amounts; and therefore that the severity of the law should be relaxed in favour of all mercantile concerns in proportion to their extent: encouragement must be given to large capitalists; and where an extensive brewery or distilleryyields an important contribution to the revenue, no strict scrutiny need be adopted in regard to the quality of the article from which such contribution is raised, provided the excise do not suffer by the fraud.

But the principles of the constitution afford no sanction to this preference, and the true interests of the country require that it should be abolished; for a tax dependent upon deception must be at best precarious, and must be, sooner or later, diminished by the irresistible diffusion of knowledge. Sound policy requires that the law should be impartially enforced in all cases; and if its penalties were extended to abuses of which it does not now take cognisance, there is no doubt that the revenue would be abundantly benefited.

Another species of fraud, to which I shall at present but briefly advert, and which has increased to so alarming an extent, that it loudly calls for the interference of government, is the adulteration of drugs and medicines.

Nine-tenths of the most potent drugs and chemical preparations used in pharmacy, are vended in a sophisticated state by dealers who would be the last to be suspected. It is well known, that of the article Peruvian bark, there is a variety of species inferior to the genuine; that too little discrimination is exercised by the collectors of this precious medicament; that it is carelessly assorted, and is frequently packed in green hides; that much of it arrives in Spain in a half-decayed state, mixed with fragments of other vegetables and various extraneous substances; and in this state is distributed throughout Europe.

But as if this were not a sufficient deterioration, the public are often served with a spurious compound of mahogany saw-dust and oak wood, ground into powder mixed with a proportion of good quinquina, and sold as genuine bark powder.

Every chemist knows that there are mills constantly at work in this metropolis, which furnish bark powder at a much cheaper rate than the substance can be procured for in its natural state. The price of the best genuine bark, upon an average, is not lower than twelve shillings the pound; but immense quantities of powder bark are supplied to the apothecaries at three or four shillings a pound.

It is also notorious that there are manufacturers of spurious rhubarb powder, ipecacuanha powder, James's powder; and other simple and compound medicines of great potency, who carry on their diabolical trade on an amazingly large scale. Indeed, the quantity of medical preparations thus sophisticated exceeds belief. Cheapness, and not genuineness and excellence, is the grand desideratum with the unprincipled dealers in drugs and medicines.

Those who are familiar with chemistry may easily convince themselves of the existence of the fraud, by subjecting to a chemical examination either spirits of hartshorn, magnesia, calcined magnesia, calomel, or any other chemical preparation in general demand.

Spirit of hartshorn is counterfeited by mixing liquid caustic ammonia with the distilled spirit of hartshorn, to increase the pungency of its odour, and to enable it to bear an addition of water.

The fraud is detected by adding spirit of wine to the sophisticated spirit; for, if no considerable coagulation ensues, the adulteration is proved. It may also be discovered by the hartshorn spirit not producing a brisk effervescence when mixed with muriatic or nitric acid.

Magnesia usually contains a portion of lime, originating from hard water being used instead of soft, in the preparation of this medicine.

To ascertain the purity of magnesia, add to a portion of it a little sulphuric acid, diluted with ten times its bulk of water. If the magnesia be completely soluble, and the solution remains transparent, it may be pronounced *pure;* but not otherwise. Or, dissolve a portion of the magnesia in muriatic acid, and add a solution of sub-carbonate of ammonia. If any lime be present, it will form a precipitate; whereas pure magnesia will remain in solution.

Calcined magnesia is seldom met with in a pure state. It may be assayed by the same tests as the common magnesia. It ought not to effervesce at all, with dilute sulphuric acid; and, if the magnesia and acid be put together into one scale of a balance, no diminution of weight should ensue on mixing them together. Calcined magnesia, however, is very seldom so pure as to be totally dissolved by diluted sulphuric acid; for a small insoluble residue generally remains, consisting chiefly of silicious earth, derived from the alkali employed in the preparation of it. The solution in sulphuric acid, when largely diluted, ought not to afford any precipitation by the addition of oxalate of ammonia.

The genuineness of calomel may be ascertained by boiling, for a few minutes, one part, with 1/32 part of muriate of ammonia in ten parts of distilled water. When carbonate of potash is added to the filtered solution, no precipitation will ensue if the calomel be pure.

Indeed, some of the most common and cheap drugs do not escape the adulterating hand of the unprincipled druggist. Syrup of buckthorn, for example, instead of being prepared from the juice of buckthorn berries, (*rhamnus catharticus,*) is made from the fruit of the blackberry bearing alder, and the dogberry tree. A mixture of the berries of the buckthorn and blackberry bearing alder, and of the dogberry tree, may be seen publicly exposed for sale by some of the venders of medicinal herbs. This abuse may be discovered by opening the berries: those of buckthorn have almost always four seeds; of the alder, two; and of the dogberry, only one. Buckthorn berries, bruised on white paper, stain it of a green colour, which the others do not.

Instead of worm-seed (*artemisia santonica,*) the seeds of tansy are frequently offered for sale, or a mixture of both.

A great many of the essential oils obtained from the more expensive spices, are frequently so much adulterated, that it is not easy to meet with such as are at all fit for use: nor are these adulterations easily discoverable. The grosser abuses, indeed, may be readily detected. Thus, if the oil be adulterated with alcohol, it will turn milky on the addition of water; if with expressed oils, alcohol will dissolve the volatile, and leave the other behind; if with oil of turpentine, on dipping a piece of paper in the mixture, and drying it with a gentle heat, the turpentine will be betrayed by its smell. The more subtile artists, however, have contrived other methods of sophistication, which elude all trials. And as all volatile oils agree in the general properties of solubility in spirit of wine, and volatility in the heat of boiling water, &c. it is plain that they may be variously mixed with each other, or the dearer sophisticated with the cheaper, without any possibility of discovering the abuse by any of the before-mentioned trials. Perfumers assert that the smell and taste are the only certain tests of which the nature of the thing will admit. For example, if a bark should have in every respect the appearance of good cinnamon, and should be proved indisputably to be the genuine bark of the cinnamon tree; yet if it want the cinnamon flavour, or has it but in a low degree, we reject it: and the case is the same with the essential oil of cinnamon. It is only from use and habit, or comparisons with specimens of known quality, that we can judge of the goodness, either of the drugs themselves, or of their oils.

Most of the arrow-root, the fecula of the Maranta arudinacea, sold by druggists, is a mixture of potatoe starch and arrow-root.

The same system of adulteration extends to articles used in various trades and manufactures. For instance, linen tape, and various other household commodities of that kind, instead of being manufactured of linen thread only, are made up of linen and cotton. Colours for painting, not only those used by artists, such as ultramarine, carmine, and lake; Antwerp blue,chrome yellow, and Indian ink; but also the coarser colours used by the common house-painter are more or less adulterated. Thus, of the latter kind, white lead is mixed with carbonate or sulphate of barytes; vermilion with red lead.

Soap used in house-keeping is frequently adulterated with a considerable portion of fine white clay, brought from St. Stephens, in Cornwall. In the manufacture of printing paper, a large quantity of plaster of Paris is added to the paper stuff, to increase the weight of the manufactured article. The selvage of cloth is often dyed with a permanent colour, and artfully stitched to the edge of cloth dyed with a fugitive dye. The frauds committed in the tanning of skins, and in the manufacture of cutlery and jewelry, exceed belief.

The object of all unprincipled modern manufacturers seems to be the sparing of their time and labour as much as possible, and to increase the quantity of the articles they produce, without much regard to their quality. The ingenuity and perseverance of self-interest is proof against prohibitions, and contrives to elude the vigilance of the most active government.

The eager and insatiable thirst for gain, which seems to be a leading characteristic of the times, calls into action every human faculty, and gives an irresistible impulse to the power of invention; and where lucre becomes the reigning principle, the possible sacrifice of even a fellow creature's life is a secondary consideration. In reference to the deterioration of almost all the necessaries and comforts of existence, it may be justly observed, in a civil as well as a religious sense, that "*in the midst of life we are in death.*"

REMARKS

ON THE

Effect of different Kinds of Waters

IN THEIR APPLICATION TO

DOMESTIC ECONOMY AND THE ARTS;

AND

METHODS OF ASCERTAINING THEIR PURITY.

It requires not much reflection to become convinced that the waters which issue from the recesses of the earth, and form springs, wells, rivers, or lakes, often materially differ from each other in their taste and other obvious properties. There are few people who have not observed a difference in the waters used for domestic purposes and in the arts; and the distinctions of*hard* and *soft* water are familiar to every body.

Water perfectly pure is scarcely ever met with in nature.

It must also be obvious, that the health and comfort of families, and the conveniences of domestic life, are materially affected by the supply of good and wholesome water. Hence a knowledge of the quality and salubrity of the different kinds of waters employed in the common concerns of life, on account of the abundant daily use we make of them in the preparation of food, is unquestionably an object of considerable importance, and demands our attention.

The effects produced by the foreign matters which water may contain, are more considerable, and of greater importance, than might at first be imagined. It cannot be denied, that such waters as are *hard*, or loaded with earthy matter, have a decided effect upon some important functions of the human body. They increase the distressing symptoms under which those persons labour who are afflicted with what is commonly called gravel complaints; and many other ailments might be named, that are always aggravated by the use of waters abounding in saline and earthy substances.

The purity of the waters employed in some of the arts and manufactures, is an object of not less consequence. In the process of brewing malt liquors, soft water is preferable to hard. Every brewer knows that the largest possible quantity of the extractive matter of the malt is obtained in the least possible time, and at the smallest cost, by means of soft water.

In the art of the dyer, hard water not only opposes the solution of several dye stuffs, but it also alters the natural tints of some delicate colours; whilst in others again it precipitates the earthy and saline matters with which it is impregnated, into the delicate fibres of the stuff, and thus impedes the softness and brilliancy of the dye.

The bleacher cannot use with advantage waters impregnated with earthy salts; and a minute portion of iron imparts to the cloth a yellowish hue.

To the manufacturer of painters' colours, water as pure as possible is absolutely essential for the successful preparation of several delicate pigments. Carmine, madder lake, ultramarine, and Indian yellow, cannot be prepared without perfectly pure water.

For the steeping or raiting of flax, soft water is absolutely necessary; in hard water the flax may be immersed for months, till its texture be injured, and still the ligneous matter will not be decomposed, and the fibres properly separated.

In the culinary art, the effects of water more or less pure are likewise obvious. Good and pure water softens the fibres of animal and vegetable matters more readily than such as is called*hard*. Every cook knows that dry or ripe pease, and other farinaceous seeds, cannot *readily* be boiled soft in hard water; because the farina of the seed is not perfectly soluble in water loaded with earthy salts.

Green esculent vegetable substances are more tender when boiled in soft water than in hard water; although hard water imparts to them a better colour. The effects of hard and soft water may be easily shown in the following manner.

EXPERIMENT.

Let two separate portions of tea-leaves be macerated, by precisely the same processes, in circumstances all alike, in similar and separate vessels, the one containing hard and the other soft water, either hot or cold, the infusion made with the soft water will have by far the strongest taste, although it possesses less colour than the infusion made with the hard water. It will strike a more intense black with a solution of sulphate of iron, and afford a more abundant precipitate, with a solution of animal jelly, which at once shews that soft water has extracted more tanning matter, and more gallic acid, from the tea-leaves, than could be obtained from them under like circumstances by means of hard water.

Many animals which are accustomed to drink soft water, refuse hard water. Horses in particular prefer the former. Pigeons refuse hard water when they have been accustomed to soft water.

CHARACTERS OF GOOD WATER.

A good criterion of the purity of water fit for domestic purposes, is its softness. This quality is at once obvious by the touch, if we only wash our hands in it with soap. Good water should be beautifully transparent; a slight opacity indicates extraneous matter. To judge of the perfect transparency of water, a quantity of it should be put into a deep glass vessel, the larger the better, so that we can look down perpendicularly into a considerable mass of the fluid; we may then readily discover the slightest degree of muddiness much better than if the water be viewed through the glass placed between the eye and the light. It should be perfectly colourless, devoid of odour, and its taste soft and agreeable. It should send out air-bubbles when poured from one vessel into another; it should boil pulse soft, and form with soap an uniform opaline fluid, which does not separate after standing for several hours.

It is to the presence of common air and carbonic acid gas that common water owes its taste, and many of the good effects which it produces on animals and vegetables. Spring water, which contains more air, has a more lively taste than river water.

Hence the insipid or vapid taste of newly boiled water, from which these gases are expelled: fish cannot live in water deprived of those elastic fluids.

100 cubic inches of the New River water, with which part of this metropolis is supplied, contains 2,25 of carbonic acid, and 1,25 of common air. The water of the river Thames contains rather a larger quantity of common air, and a smaller portion of carbonic acid.

If water not fully saturated with common air be agitated with this elastic fluid, a portion of the air is absorbed; but the two chief constituent gases of the atmosphere, the oxygen and nitrogen, are not equally affected, the former being absorbed in preference to the latter.

According to Mr. Dalton, in agitating water with atmospheric air, consisting of 79 of nitrogen, and 21 of oxygen, the water absorbs 1/64 of 79/100 nitrogen gas = 1,234, and 1/27 of 21/100 oxygen gas = 778, amounting in all to 2,012.

Water is freed from foreign matter by distillation: and for any chemical process in which accuracy is requisite, distilled water must be used.

Hard waters may, in general, be cured in part, by dropping into them a solution of sub-carbonate of potash; or, if the hardness be owing only to the presence of super-carbonate of lime, mere boiling will greatly remedy the defect; part of the carbonic acid flies off, and a neutral carbonate of lime falls down to the bottom; it may then be used for washing, scarcely curdling soap. But if the hardness be owing in part to sulphate of lime, boiling does not soften it at all.

When spring water is used for washing, it is advantageous to leave it for some time exposed to the open air in a reservoir with a large surface. Part of the carbonic acid becomes thus dissipated, and part of the carbonate of lime falls to the bottom. Mr. Dalton has observed that the more any spring is drawn from, the softer the water becomes.

CHEMICAL CONSTITUTION OF THE WATERS USED IN DOMESTIC ECONOMY AND THE ARTS.

Rain Water,

Collected with every precaution as it descends from the clouds, and at a distance from large towns, or any other object capable of impregnating the atmosphere with foreign matters, approaches more nearly to a state of purity than perhaps any other natural water. Even collected under these circumstances, however, it invariably contains a portion of common air and carbonic acid gas. The specific gravity of rain water scarcely differs from that of distilled water; and from the minute portions of the foreign ingredients which it generally contains, it is very *soft*, and admirably adapted for many culinary purposes, and various processes in different manufactures and the arts.

Fresh-fallen ***snow***, melted without the contact of air, appears to be nearly free from air. Gay-Lussac and Humboldt, however, affirm, that it contains nearly the usual proportion of air.

Water from melted *ice* does not contain so much air. *Dew* has been supposed to be saturated with air.

Snow water has long laid under the imputation of occasioning those strumous swellings in the neck which deform the inhabitants of many of the Alpine vallies; but this opinion is not supported by any well-authenticated indisputable facts, and is rendered still more improbable, if not entirely overturned, by the frequency of the disease in Sumatra, where ice and snow are never seen.

In high northern latitudes, thawed snow forms the constant drink of the inhabitants during winter; and the vast masses of ice which float on the polar seas, afford an abundant supply of fresh water to the mariner.

Spring Water,

Includes well-water and all others that arise from some depth below the surface of the earth, and which are used at the fountain-head, or at least before they have run any considerable distance exposed to the air. Indeed, springs may be considered as rain water which has passed through the fissures of the earth, and, having accumulated at the bottom of declivities, rises again to the surface forming springs and wells. As wells take their origin at some depth from the surface, and below the influence of the external atmosphere, their temperature is in general pretty uniform during every vicissitude of season, and always several degrees lower than the atmosphere. They differ from one another according to the nature of the strata through which they issue; for though the ingredients usually existing in them are in such minute quantities as to impart to the water no striking properties, and do not render it unfit for common purposes, yet they modify its nature very considerably. Hence the water of some springs is said to be *hard*, of others *soft*, some *sweet*, others *brackish*, according to the nature and degree of the inpregnating ingredients.

Common springs are insensibly changed into mineral or medicinal springs, as their foreign contents become larger or more unusual; or, in some instances, they derive medicinal celebrity from the absence of those ingredients usually occurring in spring-water; as, for example, is the case with the Malvern spring, which is nearly pure water.

Almost all spring-waters possess the property termed *hardness* in a greater or less degree; a property which depends chiefly upon the presence of super-carbonate, or of sulphate of lime, or of both; and the quantity of these earthy salts varies very considerably in different instances. Mr. Dalton has shewn that one grain of sulphate of lime, contained in 2000 grains of water, converts it into the hardest spring water that is commonly met with.

The waters of deep wells are usually much harder than those of springs which overflow the mouth of the well; but there are some exceptions to this rule.

The purest springs are those which occur in primitive rocks, or beds of gravel, or filter through sand or silicious strata. In general, large springs are purer than small ones: and our old wells contain finer water than those that are new, as the soluble parts through which the water filters in channels under ground become gradually washed away.

River Water,

Is a term applied to every running stream or rivulet exposed to the air, and always flowing in an open channel. It is formed of spring water, which, by exposure, becomes more pure, and of running land or surface water, which, although turbid from particles of the alluvial soil suspended in it, is otherwise very pure. It is purest when it runs over a gravelly or rocky bed, and when its course is swift. It is generally soft, and more free from earthy salts than spring water; but it usually contains less common air and carbonic acid gas; for, by the agitation of a long current, and exposed to the temperature of the atmosphere, part of its carbonic acid gas is disengaged, and the lime held in solution by it is in part precipitated, the loss of which contributes to the softness of the water. Its specific gravity thereby becomes less, the taste not so harsh, but less fresh and agreeable; and out of a hard spring is often made a stream of sufficient purity for most of the purposes where a soft water is required.

The water called in this metropolis *New River Water*, contains a minute portion of muriate of lime, carbonate of lime, and muriate of soda.

Some streams, however, that arise from clean silicious beds, and flow in a sandy or stony channel, are from the outset remarkably pure; such as the mountain lakes and rivulets in the rocky districts of Wales, the source of the beautiful waters of the Dee, and numberless other rivers that flow through the hollow of every valley. Switzerland has long been celebrated for the purity and excellence of its waters, which pour in copious streams from the mountains, and give rise to the finest rivers in Europe.

Some rivers, however, that do not take their rise from a rocky soil, and are indeed at first considerably charged with foreign matter, during a long course, even over a richly cultivated plain, become remarkably pure as to saline contents; but often fouled with mud containing much animal and vegetable matter, which are rather suspended than held in true solution. Such is the water of the river Thames, which, taken up at London at low water mark, is very soft and good; and, after rest, it contains but a very small portion of any thing that could prove pernicious, or impede any manufacture. It is also excellently fitted for sea-store; but it then undergoes a remarkable spontaneous change, when preserved in wooden casks. No water carried to sea becomes putrid sooner than that of the Thames. But the mode now adopted in the navy of substituting iron tanks for wooden casks, tends greatly to obviate this disadvantage.

Whoever will consider the situation of the Thames, and the immense population along its banks for so many miles, must at once perceive the prodigious accumulation of animal matters of all kinds, which by means of the common sewers constantly make their way into it. These matters are, no doubt, in part the cause of the putrefaction which it is well known to undergo at sea, and of the carburetted and sulphuretted hydrogen gases which are evolved from it. When a wooden cask is opened, after being kept a month or two, a quantity of carburetted and sulphuretted hydrogen escapes, and the water is so black and offensive as scarcely to be borne. Upon racking it off, however, into large earthen vessels, and exposing it to the air, it gradually deposits a quantity of black slimy mud, becomes clear as crystal, and remarkably sweet and palatable.

It might, at first sight, be expected that the water of the Thames, after having received all the contents of the sewers, drains, and water courses, of a large town, should acquire thereby such impregnation with foreign matters, as to become very impure; but it appears, from the most accurate experiments that have been made, that those kinds of impurities have no perceptible influence on the salubrious quality of a mass of water so immense, and constantly kept in motion by the action of the tides.

Some traces of animal matter may, however, be detected in the water of the Thames; for if nitrate of lead be dropped into it, "you will find that it becomes milky, and that a white powder falls to the bottom, which dissolves without effervescence in nitric acid. It is, therefore, (says Dr. Thomson) a combination of oxide of lead with some animal matter."

SUBSTANCES USUALLY CONTAINED IN COMMON WATER, AND TESTS BY WHICH THEY ARE DETECTED.

To acquire a knowledge of the general nature of common water, it is only necessary to add to it a few chemical tests, which will quickly indicate the presence or absence of the substances that may be expected.

Almost the only salts contained in common waters are the carbonates, sulphates, and muriates of soda, lime, and magnesia; and sometimes a very minute portion of iron may also be detected in them.

EXPERIMENT.

Fill a wine-glass with distilled water, and add to it a few drops of a solution of soap in alcohol, the water will remain transparent.

This test is employed for ascertaining the presence of earthy salts in waters. Hence it produces no change when mingled with distilled or perfectly pure water; but when added to water containing earthy salts, a white flocculent matter becomes separated, which speedily collects on the surface of the fluid. Now, from the quantity of flocculent matter produced, in equal quantities of water submitted to the test, a tolerable notion may be formed of the degrees of hardness of different kinds of water, at least so far as regards the fitness of the water for the ordinary purposes of domestic economy. This may be rendered obvious in the following manner.

EXPERIMENT.

Fill a number of wine-glasses with different kinds of pump or well water, and let fall into each glass a few drops of the solution of soap in alcohol. A turbidness will instantly ensue, and a flocculent matter collect on the surface of the fluid, if the mixture be left undisturbed. The quantity of flocculent matter will be in the ratio of the quantity of earthy salts contained in the water.

It is obvious that the action of this test is not discriminative, with regard to the chemical nature of the earthy salt present in the water. It serves only to indicate the *presence* or *absence* of those kinds of substances which occasion that quality in water which is usually called *hardness*, and which is always owing to salts with an earthy base.

If we wish to know the nature of the different acids and earths contained in the water, the following test may be employed.

EXPERIMENT.

Add about twenty drops of a solution of oxalate of ammonia, to half a wine-glass of the water; if a white precipitate ensues, we conclude that the water contains lime.

By means of this test, one grain of lime may be detected in 24,250 of water.

If this test occasion a white precipitate in water taken fresh from the pump or spring, and not after the water has been boiled and suffered to grow cold, the lime is dissolved in the water by an excess of carbonic acid; and if it continues to produce a precipitate in the water which has been concentrated by boiling, we then are sure that the lime is combined with a fixed acid.

EXPERIMENT.

To detect the presence of iron, add to a wine-glassful of the water a few drops of an infusion of nut-galls; or better, suffer a nut-gall to be suspended in it for twenty-four hours, which will cause the water to acquire a blueish black colour, if iron be present.

EXPERIMENT.

Add a few grains of muriate of barytes, to half a wine-glass of the water to be examined; if it produces a turbidness which does not disappear by the admixture of a few drops of muriatic acid, the presence of sulphuric acid is rendered obvious.

EXPERIMENT.

If a few drops of a solution of nitrate of silver occasions a milkiness with the water, which vanishes again by the copious addition of liquid ammonia, we have reason to believe that the water contains a salt, one of the constituent parts of which is muriatic acid.

EXPERIMENT.

If lime water or barytic water occasions a precipitate which again vanishes by the admixture of muriatic acid, then carbonic acid is present in the water.

EXPERIMENT.

If a solution of phosphate of soda produces a milkiness with the water, after a previous addition to it of a similar quantity of neutral carbonate of ammonia, we may then expect magnesia. The application of this test is best made in the following manner:

Concentrate a quantity of the water to be examined to about 1/20 part of its bulk, and drop into about half a wine-glassful, about five grains of neutral carbonate of ammonia. No magnesia becomes yet precipitated if this earth be present; but on adding a like quantity of phosphate of soda, the magnesia falls down, as an insoluble salt. It is essential that the carbonate of ammonia be neutral.

This test was first pointed out by Dr. Wollaston.

The presence of oxygen gas loosely combined in water may readily be discovered in the following manner.

EXPERIMENT.

Fill a vial with water, and add to it a small quantity of green sulphate of iron. If the water be entirely free of oxygen, and if the vessel be well stopped and completely filled, the solution is transparent; but if otherwise, it soon becomes slightly turbid, from the oxide of iron attracting the oxygen, and a small portion of it, in this more highly oxidated state, leaving the acid and being precipitated. Or, according to a method pointed out by Driessen, the water is to be boiled for two hours in a flask filled with it, and immersed in a vessel of water kept boiling, with the mouth of the flask under the surface of the water: it is to be inverted in quicksilver, taking care that no air-bubble adheres to the side of the flask, and being tinged with infusion of litmus, alittle nitrous gas is to be introduced: if the oxygen gas has been sufficiently expelled from the water, the purple colour of the litmus does not change; while, if oxygen be present, it immediately becomes red.

If we examine the different waters which are used for the ordinary purposes of life, and judge of them by the above tests, we shall find them to differ considerably from each other. Some contain a large quantity of saline and earthy matters, whilst others are nearly pure. The differences are produced by the great solvent power which water exercises upon most substances. Wells should never be lined with bricks, which render soft water hard; or, if bricks be employed, they should be bedded in and covered with cement.

METHOD OF ASCERTAINING THE RELATIVE QUANTITY OF EACH OF THE DIFFERENT SUBSTANCES USUALLY CONTAINED IN COMMON WATER.

To ascertain the quantity of earthy and saline matter contained in water, the following is the most simple and easy method.

EXPERIMENT.

Put any measured quantity of the water into a platina, or silver evaporating basin, the weight of which is known, and evaporate the water upon a steam bath, at a temperature of about 180°, nearly to dryness; and, lastly, remove the basin to a sand bath, and let the mass be evaporated to perfect dryness. The weight of the platina basin being already known, we have only to weigh it carefully. When the solid saline contents of the water is attached to it, the increase of weight gives the quantity of solid matter contained in a given quantity of the water.

EXPERIMENT.

Pour upon the saline contents a quantity of distilled water equal to that in which the obtained salts were originally dissolved. If the whole saline matter become dissolved in this water, there is reason to believe that the saline matter has not been altered during the evaporation of the water. But if a portion remain undissolved, as is usually the case, then we may conclude that some of the salts have mutually decomposed each other, when brought into a concentrated state by the evaporation, and that salts have been formed which did not originally exist in the water before its evaporation.

We have already mentioned that almost the only salts contained in common waters, are the carbonates, sulphates, and muriates, of soda, lime, and magnesia; and sometimes a very minute portion of iron. Having determined the different acids and bases present, in the manner stated, we may easily ascertain the relative weight of each.

The following formula suggested by Dr. Murray, is fully as accurate a means of analysing waters as any other, and it is easy of execution. The weight of the saline ingredients of a given quantity of water being determined, we may proceed to the accurate analysis of it in the following manner.

EXPERIMENT.

Measure out a determinate volume of the water (as 500 or 1000 cubic inches,) and evaporate it gradually, in an unglazed open vessel defended from dust, to one third of its original bulk; then divide this evaporated liquid into three equal portions.

EXPERIMENT.

Drop into the first portion, muriate of barytes; wash the precipitate, collect it, dry it at a red heat upon platina foil, and weigh it; digest it in nitric acid, dry it, and weigh it again. The loss of weight indicates the quantity of carbonate of barytes which the precipitate contained. The residual weight is sulphate of barytes; the carbonic acid in the water is equivalent to 0,22 of the weight of the carbonate of barytes; the sulphuric acid to 0,339 of the weight of the sulphate of barytes.

EXPERIMENT.

Precipitate the second portion of the concentrated water, by the addition of nitrate of silver; wash the precipitate, dry it, and fuse it on a piece of foil platina, previously weighed. By weighing the foil containing the fused chloride of silver, the weight of the precipitate may be ascertained. The fourth part of this weight is equivalent to the weight of the muriatic acid contained in the portion of water precipitated.

EXPERIMENT.

Precipitate the third portion of the water by the addition of oxalate of ammonia; wash and dry the precipitate; expose it to a red heat, on a platina foil, or in a capsule of platina; pour on it some dilute sulphuric acid; digest for some time, then evaporate to dryness, expose the capsule to a pretty strong heat, and, lastly, weigh the sulphate of lime thus produced: 0.453 of its weight indicate the quantity of lime in the portion of water precipitated.

EXPERIMENT.

Add to the same third portion of the water thus freed from lime, a portion of a solution of neutral carbonate of ammonia, and then add phosphoric acid, drop by drop, as long as any precipitate falls down. Wash the precipitate, dry it, and expose it to a red heat in a platina capsule: it is phosphate of magnesia. 0.357 of the weight of this salt is equivalent to the weight of the magnesia contained in the water.

EXPERIMENT.

If the water contain a minute portion of iron, a quantity of it equal to one of the three preceding portions, must be taken and mixed with a solution of benzoate of ammonia. The precipitate being washed, dried, and exposed to a red heat, and weighed, nine-tenths of its weight indicate the weight of protoxide of iron contained in the water.

In this manner the quantity of all the substances contained in the water will be ascertained, except there be any soda. To know the amount of it, the following method, pointed out by Dr. Murray, answers very well.

EXPERIMENT.

Evaporate a portion of the water to one third of its bulk. Precipitate the carbonic and sulphuric acids by the addition of muriate of barytes, taking care not to add any excess of the tests.

Precipitate the lime by oxalate of ammonia, and the magnesia by carbonate of ammonia and phosphoric acid. Then evaporate the liquid thus treated to dryness. A quantity of common salt will remain: let this be exposed to a red heat; 0.4 of its weight indicate the sodium contained in the bulk of water employed; and 0.4 sodium are equivalent to 0.53 of soda.

It seems hardly requisite to mention some other substances that occasionally make their appearance in the waters used for domestic purposes. A fine divided sand is a common constituent, which is easily obtained in a separate state. We have only to evaporate a portion of the water to dryness, and redissolve the saline residue in distilled water. The silicious sand remains undissolved, and betrays itself by its insolubility in acids, and its easy fusibility into a transparant glass, with soda, before the blow-pipe.

DELETERIOUS EFFECTS OF KEEPING WATER FOR DOMESTIC ECONOMY IN LEADEN RESERVOIRS.

The deleterious effect of lead, when taken into the stomach, is at present so universally known, that it is quite unnecessary to adduce any argument in proof of its dangerous tendency.

The ancients were, upwards of 2000 years ago, as well aware of the pernicious quality of this metal as we are at the present day; and indeed they appeared to have been much more apprehensive of its effects, and scrupulous in the application of it to purposes of domestic economy.

Their precautions may have been occasionally carried to an unnecessary length. This was the natural consequence of the imperfect state of experimental knowledge at that period. When men were unable to detect the poisonous matters—to be over scrupulous in the use of such water, was an error on the right side.

The moderns, on the other hand, in part, perhaps, from an ill-founded confidence, and inattention to a careful and continued examination of its effects, have fallen into an opposite error.

There can be no doubt that the mode of preserving water intended for food or drink in leaden reservoirs, is exceedingly improper; and although pure water exercises no sensible action upon metallic lead, provided air be excluded, the metal is certainly acted on by the water when air is admitted: this effect is so obvious, that it cannot escape the notice of the least attentive observer.

The white line which may be seen at the surface of the water preserved in leaden cisterns, where the metal touches the water and where the air is admitted, is a carbonate of lead, formed at the expense of the metal. This substance, when taken into the stomach, is highly deleterious to health. This was the reason which induced the ancients to condemn leaden pipes for the conveyance of water; it having been remarked that persons who swallowed the sediment of such water, became affected with disorders of the bowels.

Leaden water reservoirs were condemned in ancient times by Hyppocrates, Galen, and Vitruvius, as dangerous: in addition to which, we may depend on the observations of Van Swieten, Tronchin, and others, who have quoted numerous unhappy examples of whole families poisoned by water which had remained in reservoirs of lead. Dr. Johnston, Dr. Percival, Sir George Baker, and Dr. Lamb, have likewise recorded numerous instances where dangerous diseases ensued from the use of water impregnated with lead.

Different potable waters have unequal solvent powers on this metal. In some places the use of leaden pumps has been discontinued, from the expense entailed upon the proprietors by the constant want of repair. Dr. Lamb states an instance where the proprietor of a well ordered his plumber to make the lead of a pump of double the thickness of the metal usually employed for pumps, to save the charge of repairs; because he had observed that the water was so hard, as he called it, that it corroded the lead very soon.

The following instance is related by Sir George Baker:

"A gentleman was the father of a numerous offspring, having had one-and-twenty children, of whom eight died young, and thirteen survived their parents. During their infancy, and indeed*until they had quitted the place of their usual residence, they were all remarkably* *unhealthy*; being particularly subject to disorders of the stomach and bowels. The father, during many years, was paralytic; the mother, for a long time, was subject to colics and bilious obstructions.

"After the death of the parents, the family sold the house which they had so long inhabited. The purchaser found it necessary to repair the pump. This was made of lead; which, upon examination was found to be so corroded, that several perforations were observed in the cylinder, in which the bucket plays; and the cistern in the upper part was reduced to the thinness of common brown paper, and was full of holes, like a sieve."

I have myself seen numerous instances where leaden cisterns have been completely corroded by the action of water with which they were in contact: and there is, perhaps, not a plumber who cannot give testimony of having experienced numerous similar instances in the practice of his trade.

I have been frequently called upon to examine leaden cisterns, which had become leaky on account of the action of the water which they contained; and I could adduce an instance of a legal controversy having taken place to settle the disputes between the proprietors of an estate and a plumber, originating from a similar cause—the plumber being accused of having furnished a faulty reservoir; whereas the case was proved to be owing to the chemical action of the water on the lead. Water containing a large quantity of common air and carbonic acid gas, always acts very sensibly on metallic lead.

Water, which has no sensible action, in its natural state, upon lead, may acquire the capability of acting on it by heterogeneous matter, which it may accidentally receive. Numerous instances have shewn that vegetable matter, such as leaves, falling into leaden cisterns filled with water, imparted to the water a considerable solvent power of action on the lead, which, in its natural state it did not possess. Hence the necessity of keeping leaden cisterns clean; and this is the more necessary, as their situations expose them to accidental impurities. The noted saturnine colic of Amsterdam, described by Tronchin, originated from such a circumstance; as also the case related by Van Swieten, of a whole family afflicted with the same complaint, from such a cistern. And it is highly probable that the case of disease recorded by Dr. Duncan, proceeded more from some foulness in the cistern, than from the solvent power of the water. In this instance the officers of the packet boat used water for their drink and cooking out of a leaden cistern, whilst the sailors used the water taken from the same source, except that theirs was kept in wooden vessels. The consequence was, that all the officers were seized with the colic, and all the men continued healthy.

The carelessness of the bulk of mankind, Dr. Lambe very justly observes, to these things, "is so great, that to repeat them again and again cannot be wholly useless."

Although the great majority of persons who daily use water kept in leaden cisterns receive no sensible injury, yet the apparent salubrity must be ascribed to the great slowness of its operation, and the minuteness of the dose taken, the effects of which become modified by different causes and different constitutions, and according to the predisposition to diseases inherent in different individuals. The supposed security of the multitude who use the water with impunity, amounts to no more than presumption, in favour of any individual, which may or may not be confirmed by experience.

Independent of the morbid susceptibility of impressions which distinguish certain habits, there is, besides, much variety in the original constitution of the human frame, of which we are totally ignorant.

"The susceptibility or proneness to disease of each individual, must be esteemed peculiar to himself. Confiding to the experience of others is a ground of security which may prove fallacious; and the danger can with certainty be obviated only by avoiding its source. And considering the various and complicated changes of the human frame, under different circumstances and at different ages, it is neither impossible nor improbable that the substances taken into the system at one period, and even for a series of years, with apparent impunity may, notwithstanding, at another period, be eventually the occasion of disease and of death.

"The experience of a single person, or of many persons, however numerous, is quite incompetent to the decision of a question of this nature.

"The pernicious effects of an intemperate use of spiritous liquors is not less certain because we often see habitual drunkards enjoy a state of good health, and arrive at old age: and the same may be said of individuals who indulge in vices of all kinds, evidently destructive to life; many of whom, in spite of their bad habits, attain to a vigorous old age."

In confirmation of these remarks, we adduce the following account of the effect of water contaminated by lead, given by Sir G. Baker:

"The most remarkable case on the subject that now occurs to my memory, is that of Lord Ashburnham's family, in Sussex; to which, spring water was supplied, from a considerable distance, in leaden pipes. In consequence, his Lordship's servants were every year tormented with colic, and some of them died. An eminent physician, of Battle, who corresponded with me on the subject, sent up some gallons of that water, which were analysed by Dr. Higgins, who reported that the water had contained more than the common quantity of carbonic acid; and that he found in it lead in solution, which he attributed to the carbonic acid. In consequence of this, Lord Ashburnham substituted wooden for leaden pipes; and from that time his family have had no particular complaints in their bowels."

Richmond, Sept. 27, 1802.

METHOD OF DETECTING LEAD, WHEN CONTAINED IN WATER.

One of the most delicate tests for detecting lead, is water impregnated with sulphuretted hydrogen gas, which instantly imparts to the fluid containing the minutest quantity of lead, a brown or blackish tinge.

This test is so delicate that distilled water, when condensed by a leaden pipe in a still tub, is affected by it. To shew the action of this test, the following experiments will serve.

EXPERIMENT.

Pour into a wine-glass containing distilled water, an equal quantity of water impregnated with sulphuretted hydrogen gas: no change will take place; but if a 1/4 of a grain of acetate of lead (sugar of lead of commerce), or any other preparation of lead, be added, the mixture will instantly turn brown and dark-coloured.

To apply this test, one part of the suspected water need merely to be mingled with a like quantity of water impregnated with sulphuretted hydrogen. Or better, a larger quantity, a gallon for example, of the water may be concentrated by evaporation to about half a pint, and then submitted to the action of the test.

Another and more efficient mode of applying this test, is, to pass a current of sulphuretted hydrogen gas through the suspected water in the following manner.

EXPERIMENT.

Take a bottle (*a*) or Florence flask, adapt to the mouth of it a cork furnished with a glass tube (*b*), bent at right angles; let one leg of the tube be immersed in the vial (*c*) containing the water to be examined; as shewn in the following sketch. Then take one part of sulphuret of antimony of commerce, break it into pieces of half the size of split pease, put it into the flask, and pour upon it four parts of common concentrated muriatic acid (spirit of salt of commerce). Sulphuretted hydrogen gas will become disengaged from the materials in abundance, and pass through the water in the vial (*c*). Let the extrication of the gas be continued for about five minutes; and if the minutest quantity of lead be present, the water will acquire a dark-brown or blackish tinge. The extrication of the gas is facilitated by the application of a gentle heat.

The action of the sulphuretted hydrogen test, when applied in this manner, is astonishingly great; for one part of acetate of lead may be detected by means of it, in 20000 parts of water.

Another test for readily detecting lead in water, is sulphuretted chyazate of potash, first pointed out as such by Mr. Porret. A few drops of this re-agent, added to water containing lead, occasion a white precipitate, consisting of small brilliant scales of a considerable lustre.

Sulphate of potash, or sulphate of soda, is likewise a very delicate test for detecting minute portions of lead. Dr. Thomson discovered, by means of it, one part of lead in 100000 parts of water; and this acute Philosopher considers it as the most unequivocal test of lead that we possess. Dr. Thomson remarks that "no other precipitate can well be confounded with it, except sulphate of barytes; and there is no probability of the presence of barytes existing in common water."

Carbonate of potash, or carbonate of soda, may also be used as agents to detect the presence of lead. By means of these salts Dr. Thomson was enabled to detect the presence of a smaller quantity of lead in distilled water, than by the action of sulphuretted hydrogen. But the reader must here be told, that the use of these tests cannot be entrusted to an unskilful hand; because the alkaline carbonates throw down also lime and magnesia, two substances which are frequently found in common water; the former tests, namely, water impregnated with sulphuretted hydrogen gas, and nascent sulphuretted hydrogen, are therefore preferable.

It is absolutely essential that the water impregnated with sulphuretted hydrogen, when employed as a test for detecting very minute quantities of lead, be fresh prepared; and if sulphate of potash, or sulphate of soda, be used as tests, they should be perfectly pure. Sulphate of potash is preferable to sulphate of soda. It is likewise advisable to act with these tests upon water concentrated by boiling. The water to which the test has been added does sometimes appear not to undergo any change, at first; it is therefore necessary to suffer the mixture to stand for a few hours; after which time the action of the test will be more evident. Mr. Silvester has proposed gallic acid as a delicate test for detecting lead.

Adulteration of Wine.

It is sufficiently obvious, that few of those commodities, which are the objects of commerce, are adulterated to a greater extent than wine. All persons moderately conversant with the subject, are aware, that a portion of alum is added to young and meagre red wines, for the purpose of brightening their colour; that Brazil wood, or the husks of elderberries and bilberries,are employed to impart a deep rich purple tint to red Port of a pale, faint colour; that gypsum is used to render cloudy white wines transparent; that an additional astringency is imparted to immature red wines by means of oak-wood sawdust, and the husks of filberts; and that a mixture of spoiled foreign and home-made wines is converted into the wretched compound frequently sold in this town by the name of *genuine old Port*.

Various expedients are resorted to for the purpose of communicating particular flavours to insipid wines. Thus a *nutty* flavour is produced by bitter almonds; factitious Port wine is flavoured with a tincture drawn from the seeds of raisins; and the ingredients employed to form the *bouquet* of high-flavoured wines, are sweet-brier, oris-root, clary, cherry laurel water, and elder-flowers.

The flavouring ingredients used by manufacturers, may all be purchased by those dealers in wine who are initiated in the mysteries of the trade; and even a manuscript recipe book for preparing them, and the whole mystery of managing all sorts of wines, may be obtained on payment of a considerable fee.

The sophistication of wine with substances not absolutely noxious to health, is carried to an enormous extent in this metropolis. Many thousand pipes of spoiled cyder are annually brought hither from the country, for the purpose of being converted into factitious Port wine. The art of manufacturing spurious wine is a regular trade of great extent in this metropolis.

"There is, in this city, a certain fraternity of chemical operators, who work underground in holes, caverns, and dark retirements, to conceal their mysteries from the eyes and observation of mankind. These subterraneous philosophers are daily employed in the transmutation of liquors; and by the power of magical drugs and incantations, raising under the streets of London the choicest products of the hills and valleys of France. They can squeeze Bourdeaux out of the sloe, and draw Champagne from an apple. Virgil, in that remarkable prophecy,

Incultisque ruhens pendebit sentibus uva.

The ripening grape shall hang on every thorn.

seems to have hinted at this art, which can turn a plantation of northern hedges into a vineyard. These adepts are known among one another by the name of *Wine-brewers*; and, I am afraid, do great injury, not only to her Majesty's customs, but to the bodies of many of her good subjects."

The following are a few of the recipes employed in the manufacture of spurious wine:

To make *British Port Wine.*—"Take of British grape wine, or good cyder, 4 gallons; of the juice of red beet root two quarts; brandy, two quarts; logwood 4 ounces; rhatany root, bruised, half a pound: first infuse the logwood and rhatany root in brandy, and a gallon of grape wine or cyder for one week; then strain off the liquor, and mix it with the other ingredients; keep it in a cask for a month, when it will be fit to bottle."

British Champagne.—"Take of white sugar, 8 pounds; the whitest brown sugar, 7 pounds, crystalline lemon acid, or tartaric acid, 1 ounce and a quarter, pure water, 8 gallons; white grape wine, two quarts, or perry, 4 quarts; of French brandy, 3 pints."

"Put the sugar in the water, skimming it occasionally for two hours, then pour it into a tub and dissolve in it the acid; before it is cold, add some yeast and ferment. Put it into a clean cask and add the other ingredients. The cask is then to be well bunged, and kept in a cool place for two or three months; then bottle it and keep it cool for a month longer, when it will be fit for use. If it should not be perfectly clear after standing in the cask two or three months, it should be rendered so by the use of isinglass. By adding 1 lb. of fresh or preserved strawberries, and 2 ounces of powdered cochineal, the pink *Champagne may be made*."

Southampton Port.—"Take cyder, 36 gallons; elder wine, 11 gallons; brandy, 5 gallons; damson wine, 11 gallons; mix."

The particular and separate department in this factitious wine trade, called *crusting*, consists in lining the interior surface of empty wine-bottles, in part, with a red crust of super-tartrate of potash, by suffering a saturated hot solution of this salt, coloured red with a decoction of Brazil-wood, to crystallize within them; and after this simulation of maturity is perfected, they are filled with the compound called Port wine.

Other artisans are regularly employed in staining the lower extremities of bottle-corks with a fine red colour, to appear, on being drawn, as if they had been long in contact with the wine.

The preparation of an astringent extract, to produce, from spoiled home-made and foreign wines, a "genuine old Port," by mere admixture; or to impart to a weak wine a rough austere taste, a fine colour, and a peculiar flavour; forms one branch of the business of particular wine-coopers: while the mellowing and restoring of spoiled white wines, is the sole occupation of men who are called *refiners of wine*.

We have stated that a crystalline crust is formed on the interior surface of bottles, for the purpose of misleading the unwary into a belief that the wine contained in them is of a certain age. A correspondent operation is performed on the wooden cask; the whole interior of which is stained artificially with a crystalline crust of super-tartrate of potash, artfully affixed in a manner precisely similar to that before stated. Thus the wine-merchant, after bottling off a pipe of wine, is enabled to impose on the understanding of his customers, by taking to pieces the cask, and exhibiting the beautiful dark coloured and fine crystalline crust, as an indubitable proof of the age of the wine; a practice by no means uncommon, to flatter the vanity of those who pride themselves in their acute discrimination of wines.

These and many other sophistications, which have long been practised with impunity, are considered as legitimate by those who pride themselves for their skill in the art of *managing*, or, according to the familiar phrase, *doctoring* wines. The plea alleged in exculpation of them, is, that, though deceptive, they are harmless: but even admitting this as a palliation, yet they form only one department of an art which includes other processes of a tendency absolutely criminal.

Several well-authenticated facts have convinced me that the adulteration of wine with substances deleterious to health, is certainly practised oftener than is, perhaps, suspected; and it would be easy to give some instances of very serious effects having arisen from wines contaminated with deleterious substances, were this a subject on which I meant to speak..

"On the 17th of January, the passengers by the Highflyer coach, from the north, dined, as usual, at Newark. A bottle of Port wine was ordered; on tasting which, one of the passengers observed that it had an unpleasant flavour, and begged that it might be changed. The waiter took away the bottle, poured into a fresh decanter half the wine which had been objected to, and filled it up from another bottle. This he took into the room, and the greater part was drank by the passengers, who, after the coach had set out towards Grantham, were seized with extreme sickness; one gentleman in particular, who had taken more of the wine than the others, it was thought would have died, but has since recovered. The half of the bottle of wine sent out of the passengers' room, was put aside for the purpose of mixing negus. In the evening, Mr. Bland, of Newark, went into the hotel, and drank a glass or two of wine and water. He returned home at his usual hour, and went to bed; in the middle of the night he was taken so ill, as to induce Mrs. Bland to send for his brother, an apothecary in the town; but before that gentleman arrived, he was dead. An inquest was held, and the jury, after the fullest enquiry, and the examination of the surgeons by whom the body was opened, returned a verdict of—*Died by Poison*."

The most dangerous adulteration of wine is by some preparations of lead, which possess the property of stopping the progress of acescence of wine, and also of rendering white wines, when muddy, transparent. I have good reason to state that lead is certainly employed for this purpose. The effect is very rapid; and there appears to be no other method known, of rapidly recovering ropy wines. Wine merchants persuade themselves that the minute quantity of lead employed for that purpose is perfectly harmless, and that no atom of lead remains in the wine. Chemical analysis proves the contrary; and the practice of clarifying spoiled white wines by means of lead, must be pronounced as highly deleterious.

Lead, in whatever state it be taken into the stomach, occasions terrible diseases; and wine, adulterated with the minutest quantity of it, becomes a slow poison. The merchant or dealer who practises this dangerous sophistication, adds the crime of murder to that of fraud, and deliberately scatters the seeds of disease and death among those consumers who contribute to his emolument. If to debase the current coin of the realm be denounced as a capital offence, what punishment should be awarded against a practice which converts into poison a liquor used for sacred purposes.

Dr. Watson relates, that the method of adulterating wine with lead, was at one time a common practice in Paris.

Dr. Warren states an instance of thirty-two persons having become severely ill, after drinking white wine that had been adulterated with lead. One of them died, and one became paralytic.

In Graham's Treatise on Wine-Making, under the article of *Secrets*, belonging to the mysteries of vintners, lead is recommended to prevent wine from becoming acid. The following lines are copied from Mr. Graham's work:

"*To hinder Wine from turning.*

"Put a pound of melted lead, in fair water, into your cask, pretty warm, and stop it close."

"*To soften Grey Wine.*

"Put in a little vinegar wherein litharge has been well steeped, and boil some honey, to draw out the wax. Strain it through a cloth, and put a quart of it into a tierce of wine, and this will mend it."

The ancients knew that lead rendered harsh wines milder, and preserved it from acidity, without being aware that it was pernicious: it was therefore long used with confidence; and when its effects were discovered, they were not ascribed to that metal, but to some other cause. When the Greek and Roman wine merchants wished to try whether their wine was spoiled, they immersed in it a plate of lead; if the colour of the lead were corroded, they concluded that their wine was spoiled. Wine may become accidentally impregnated with lead.

It is well known that bottles in which wine has been kept, are usually cleaned by means of shot, which by its rolling motion detaches the super-tartrate of potash from the sides of the bottles. This practice, which is generally pursued by wine-merchants, may give rise to serious consequences, as will become evident from the following case:

"A gentleman who had never in his life experienced a day's illness, and who was constantly in the habit of drinking half a bottle of Madeira wine after his dinner, was taken ill, three hours after dinner, with a severe pain in the stomach and violent bowel colic, which gradually yielded within twelve hours to the remedies prescribed by his medical adviser. The day following he drank the remainder of the same bottle of wine which was left the preceding day, and within two hours afterwards he was again seized with the most violent colliquative pains, headach, shiverings, and great pain over the whole body. His apothecary becoming suspicious that the wine he had drank might be the cause of the disease, ordered the bottle from which the wine had been decanted to be brought to him, with a view that he might examine the dregs, if any were left. The bottle happening to slip out of the hand of the servant, disclosed a row of shot wedged forcibly into the angular bent-up circumference of it. On examining the beads of shot, they crumbled into dust, the outer crust (defended by a coat of black lead with which the shot is glazed) being alone left unacted on, whilst the remainder of the metal was dissolved. The wine, therefore, had become contaminated with *lead and arsenic*, the shot being a compound of these metals, which no doubt had produced the mischief."

TEST FOR DETECTING THE DELETERIOUS ADULTERATIONS OF WINE.

A ready re-agent for detecting the presence of lead, or any other deleterious metal in wine, is known by the name of the *wine test*. It consists of water saturated with sulphuretted hydrogen gas, acidulated with muriatic acid. By adding one part of it, to two of wine, or any other liquid suspected to contain lead, a dark coloured or black precipitate will fall down, which does not disappear by an addition of muriatic acid; and this precipitate, dried and fused before the blowpipe on a piece of charcoal, yields a globule of metallic lead. This test does not precipitate iron; the muriatic acid retains iron in solution when combined with sulphuretted hydrogen; and any acid in the wine has no effect in precipitating any of the sulphur of the test liquor. Or a still more efficacious method is, to pass a current of sulphuretted hydrogen gas through the wine, in the manner described, having previously acidulated the wine with muriatic acid.

The wine test sometimes employed is prepared in the following manner:—Mix equal parts of finely powdered sulphur and of slacked quick-lime, and expose it to a red heat for twenty minutes. To thirty-six grains of this sulphuret of lime, add twenty-six grains of super-tartrate of potassa; put the mixture into an ounce bottle, and fill up the bottle with water that has been previously boiled, and suffered to cool. The liquor, after having been repeatedly shaken, and allowed to become clear, by the subsidence of the undissolved matter, may then be poured into another phial, into which about twenty drops of muriatic acid have been previously put. It is then ready for use. This test, when mingled with wine containing lead or copper, turns the wine of a dark-brown or black colour. But the mere application of sulphuretted hydrogen gas to wine, acidulated by muriatic acid, is a far more preferable mode of detecting lead in wine.

M. Vogel has lately recommended acetate of lead as a test for detecting extraneous colours in red wine. He remarks, that none of the substances that can be employed for colouring wine, such as the berries of the Vaccinium Mirtillus (bilberries), elderberries, and Campeach wood, produce with genuine red wine, a greenish grey precipitate, which is the colour that is procured by this test by means of genuine red wines.

Wine coloured with the juice of the bilberries, or elderberries, or Campeach wood, produces, with acetate of lead, a deep blue precipitate; and Brazil-wood, red saunders, and the red beet, produce a colour which is precipitated red by acetate of lead. Wine coloured by beet root is also rendered colourless by lime water; but the weakest acid brings back the colour. As the colouring matter of red wines resides in the skin of the grape, M. Vogel prepared a quantity of skins, and reduced them to powder. In this state he found that they communicated to alcohol a deep red colour: a paper stained with this colour was rendered red by acids and green by alkalies.

M. Vogel made a quantity of red wine from black grapes, for the purpose of his experiments; and this produced the genuine greyish green precipitate with acetate of lead. He also found the same coloured precipitate in two specimens of red wine, the genuineness of which could not be suspected; the one from Chateau-Marguaux, and the other from the neighbourhood of Coblentz.

SPECIFIC DIFFERENCES, AND COMPONENT PARTS OF WINE.

Every body knows that no product of the arts varies so much as wine; that different countries, and sometimes the different provinces of the same country, produce different wines. These differences, no doubt, must be attributed chiefly to the climate in which the vineyard is situated—to its culture—the quantity of sugar contained in the grape juice—the manufacture of the wine; or the mode of suffering its fermentation to be accomplished. If the grapes be gathered unripe, the wine abounds with acid; but if the fruit be gathered ripe, the wine will be rich. When the proportion of sugar in the grape is sufficient, and the fermentation complete, the wine is perfect and generous. If the quantity of sugar be too large, part of it remains undecomposed, as the fermentation is languid, and the wine is sweet and luscious; if, on the contrary, it contains, even when full ripe, only a small portion of sugar, the wine is thin and weak; and if it be bottled before the fermentation be completed, part of the sugar remains undecomposed, the fermentation will go on slowly in the bottle, and, on drawing the cork, the wine sparkles in the glass; as, for example, Champagne. Such wines are not sufficiently mature. When the must is separated from the husk of the red grape before it is fermented, the wine has little or no colour: these are called *white* wines. If, on the contrary, the husks are allowed to remain in the must while the fermentation is going on, the alcohol dissolves the colouring matter of the husks, and the wine is coloured: such are called *red* wines. Hence white wines are often prepared from red grapes, the liquor being drawn off before it has acquired the red colour; for the skin of the grape only gives the colour. Besides in these principal circumstances, wines vary much in flavour.

All wines contain one common and identical principle, from which their similar effects are produced; namely, *brandy* or *alcohol*. It is especially by the different proportions of brandy contained in wines, that they differ most from one another. When wine is distilled, the alcohol readily separates. The spirit thus obtained is well known under the name of *brandy*.

All wines contain also a free acid; hence they turn blue tincture of cabbage, red. The acid found in the greatest abundance in grape wines, is tartaric acid. Every wine contains likewise a portion of super-tartrate of potash, and extractive matter, derived from the juice of the grape. These substances deposit slowly in the vessel in which they are kept. To this is owing the improvement of wine from age. Those wines which effervesce or froth, when poured into a glass, contain also carbonic acid, to which their briskness is owing. The peculiar flavour and odour of different kinds of wines probably depend upon the presence of a *volatile oil*, so small in quantity that it cannot be separated.

EASY METHOD OF ASCERTAINING THE QUANTITY OF BRANDY CONTAINED IN VARIOUS SORTS OF WINE.

The strength of all wines depends upon the quantity of alcohol or brandy which they contain. Mr. Brande, and Gay-Lussac, have proved, by very decisive experiments, that all wines contain brandy or alcohol ready formed. The following is the process discovered by Mr. Brande, for ascertaining the quantity of spirit, or brandy, contained in different sorts of wine.

EXPERIMENT.

Add to eight parts, by measure, of the wine to be examined, one part of a concentrated solution of sub-acetate of lead: a dense insoluble precipitate will ensue; which is a combination of the test liquor with the colouring, extractive, and acid matter of the wine. Shake the mixture for a few minutes, pour the whole upon a filtre, and collect the filtered fluid. It contains the brandy or spirit, and water of the wine, together with a portion of the sub-acetate of lead. Add, in small quantities at a time, to this fluid, warm, dry, and pure sub-carbonate of potash (*not salt of tartar, or sub-carbonate of potash of commerce*), which has previously been freed from water by heat, till the last portion added remains undissolved. The brandy or spirit contained in the fluid will become separated; for the sub-carbonate of potash abstracts from it the whole of the water with which it was combined; the brandy or spirit of wine forming a distinct stratum, which floats upon the aqueous solution of the alkaline salt. If the experiment be made in a glass tube, from one-half inch to two inches in diameter, and graduated into 100 equal parts, the *per centage* of spirit, in a given quantity of wine, may be read off by mere inspection. In this manner the strength of any wine may be examined.

Tabular View, exhibiting the Per Centage of Brandy or Alcohol contained in various kinds of Wines, and other fermented Liquors.

	Proportion of Spirit per Cent. by measure.
Lissa	26,47
Ditto	24,35
Average	25,41
Raisin Wine	26,40
Ditto	25,77
Ditto	23,30
Average	25,12
Marcella	26,03

Ditto	25,05
Average	25,09
Madeira	24,42
Ditto	23,93
Ditto (Sercial)	21,40
Ditto	19,24
Average	22,27
Port	25,83
Ditto	24,29
Ditto	23,71
Ditto	23,39
Ditto	22,30
Ditto	21,40
Ditto	19,96
Average	22,96
Sherry	19,81
Ditto	19,83
Ditto	18,79
Ditto	18,25
Average	19,17
Teneriffe	19,79
Colares	19,75
Lachryma Christi	19,70
Constantia (White)	19,75
Ditto (Red)	18,92
Lisbon	18,94
Malaga (1666)	18,94
Bucellas	18,49
Red Madeira	22,30
Ditto	18,40
Average	20,35
Cape Muschat	18,25
Cape Madeira	22,94
Ditto	20,50

Ditto	18,11
Average	20,51
Grape Wine	18,11
Calcavella	19,20
Ditto	18,10
Average	18,65
Vidonia	19,25
Alba Flora	17,26
Malaga	17,26
Hermitage (White)	17,43
Roussillon	19,00
Ditto	17,20
Average	18,13

	Proportion of Spirit per Cent. by measure.
Claret	17,11
Ditto	16,32
Ditto	14,08
Ditto	12,91
Average	15,10
Malmsey Madeira	16,40
Lunel	15,52
Sheraaz	15,52
Syracuse	15,28
Sauterne	14,22
Burgundy	16,60
Ditto	15,22
Ditto	14,53
Ditto	11,95
Average	14,57
Hock	14,37
Ditto	13,00
Ditto (old in cask)	8,68

Average	12,08
Nice	14,62
Barsac	13,86
Tent	13,30
Champagne (Still)	13,80
Ditto (Sparkling)	12,80
Ditto (Red)	12,56
Ditto (ditto)	11,30
Average	12,61
Red Hermitage	12,32
Vin de Grave	13,94
Ditto	12,80
Average	13,37
Frontignac	12,79
Cote Rotie	12,32
Gooseberry Wine	11,84
Currant Wine	20,55
Orange Wine aver.	11,26
Tokay	9,88
Elder Wine	9,87
Cyder highest aver.	9,87
Ditto lowest ditto	5,21
Perry average	7,26
Mead	7,32
Ale (Burton)	8,88
Ditto (Edinburgh)	6,20
Ditto (Dorchester)	5,50
Average	6,87
Brown Stout	6,80
London Porter aver.	4,20
Do. Small Beer, do.	1,28
Brandy	53,39
Rum	53,68
Gin	51,60

Scotch Whiskey	54,32
Irish ditto	53,99

CONSTITUTION OF HOME-MADE WINES.

Besides grapes, the most valuable of the articles of which wine is made, there are a considerable number of fruits from which a vinous liquor is obtained. Of such, we have in this country the gooseberry, the currant, the elderberry, the cherry, &c. which ferment well, and affords what are called *home-made wines*.

They differ chiefly from foreign wines in containing a much larger quantity of acid. Dr. Macculloch has remarked that the acid in home-made wines is principally the malic acid; while in grape wines it is the tartaric acid.

The great deficiency in these wines, independent of the flavour, which chiefly originates, not from the juice, but from the seeds and husks of the fruits, is the excess of acid, which is but imperfectly concealed by the addition of sugar. This is owing, chiefly, as Dr. Macculloch remarks, to the tartaric acid existing in the grape juice in the state of super-tartrate of potash, which is in part decomposed during the fermentation, and the rest becomes gradually precipitated; whilst the malic acid exists in the currant and gooseberry juice in the form of malate of potash; which salt does not appear to suffer a decomposition during the fermentation of the wine; and, by its greater solubility, is retained in the wine. Hence Dr. Macculloch recommends the addition of super-tartrate of potash, in the manufacture of British wines. They also contain a much larger proportion of mucilage than wines made from grapes. The juice of the gooseberry contains some portion of tartaric acid; hence it is better suited for the production of what is called *English Champagne*, than any other fruit of this country.

Adulteration of Bread.

This is one of the sophistications of the articles of food most commonly practised in this metropolis, where the goodness of bread is estimated entirely by its whiteness. It is therefore usual to add a certain quantity of alum to the dough; this improves the look of the bread very much, and renders it whiter and firmer. Good, white, and porous bread, may certainly be manufactured from good wheaten flour alone; but to produce the degree of whiteness rendered indispensable by the caprice of the consumers in London, it is necessary (unless the very best flour is employed,) that the dough should be *bleached*; and no substance has hitherto been found to answer this purpose better than alum.

Without this salt it is impossible to make bread, from the kind of flour usually employed by the London bakers, so white, as that which is commonly sold in the metropolis.

If the alum be omitted, the bread has a slight yellowish grey hue—as may be seen in the instance of what is called *home-made bread*, of private families. Such bread remains longer moist than bread made with alum; yet it is not so light, and full of eyes, or porous, and it has also a different taste.

The quantity of alum requisite to produce the required whiteness and porosity depends entirely upon the genuineness of the flour, and the quality of the grain from which the flour is obtained. The mealman makes different sorts of flour from the same kind of grain. The best flour is mostly used by the biscuit bakers and pastry cooks, and the inferior sorts in the making of bread. The bakers' flour is very often made of the worst kinds of damaged foreign wheat, and other cereal grains mixed with them in grinding the wheat into flour. In this capital, no fewer than six distinct kinds of wheaten flour are brought into market. They are called fine flour, seconds, middlings, fine middlings, coarse middlings, and twenty-penny flour. Common garden beans, and pease, are also frequently ground up among the London bread flour.

I have been assured by several bakers, on whose testimony I can rely, that the small profit attached to the bakers' trade, and the bad quality of the flour, induces the generality of the London bakers to use alum in the making of their bread.

The smallest quantity of alum that can be employed with effect to produce a white, light, and porous bread, from an inferior kind of flour, I have my own baker's authority to state, is from three to four ounces to a sack of flour, weighing 240 pounds. The alum is either mixed well in the form of powder, with a quantity of flour previously made into a liquid paste with water, and then incorporated with the dough; or the alum is dissolved in the water employed for mixing up the whole quantity of the flour for making the dough.

Let us suppose that the baker intends to convert five bushels, or a sack of flour, into loaves with the least adulteration practised. He pours the flour into the kneading trough, and sifts it through a fine wire sieve, which makes it lie very light, and serves to separate any impurities with which the flour may be mixed. Two ounces of alum are then dissolved in about a quart of boiling water, and the solution poured into *the seasoning-tub*. Four or five pounds of salt are likewise put into the tub, and a pailful of hot-water. When this mixture has cooled down to the temperature of about 84°, three or four pints of yeast are added; the whole is mixed, strained through the seasoning sieve, emptied into a hole in the flour, and mixed up with the requisite portion of it to the consistence of a thick batter. Some dry flour is then sprinkled over the top, and it is covered up with cloths.

In this situation it is left about three hours. It gradually swells and breaks through the dry flour scattered on its surface. An additional quantity of warm water, in which one ounce of alum is dissolved, is now added, and the dough is made up into a paste as before; the whole is then covered up. In this situation it is left for a few hours.

The whole is then intimately kneaded with more water for upwards of an hour. The dough is cut into pieces with a knife, and penned to one side of the trough; some dry flour is sprinkled over it, and it is left in this state for about four hours. It is then kneaded again for half-an-hour. The dough is now cut into pieces and weighed, in order to furnish the requisite quantity for each loaf. The loaves are left in the oven about two hours and a half. When taken out, they are carefully covered up, to prevent as much as possible the loss of weight.

The following account of making a sack, of five bushels of flour into bread, is taken from Dr. P. Markham's Considerations on the Ingredients used in the Adulteration of Bread Flour, and Bread:

5 bushels of flour,

8 ounces of alum,

4 lbs. of salt,

1/2 a gallon of yeast, mixed with about

3 gallons of water.

lbs.		
The whole quantity of bread-flour obtained from the bushel of wheat, weighs		48
	lbs.	
Fine pollard	4-1/4	
Coarse pollard	4	
Bran	2-3/4	
	— — —	11
—		
The whole together		59
To which add the loss of weight in manufacturing a bushel of wheat		2
—		
Produces the original weight		61
—		

The theory of the bleaching property of alum, as manifested in the panification of an inferior kind of flour, is by no means well understood; and indeed it is really surprising that the effect should be produced by so small a quantity of that substance, two or three ounces of alum being sufficient for a sack of flour.

From experiments in which I have been employed, with the assistance of skilful bakers, I am authorised to state, that without the addition of alum, it does not appear possible to make white, light, and porous bread, such as is used in this metropolis, unless the flour be of the very best quality.

Another substance employed by fraudulent bakers, is subcarbonate of ammonia. With this salt, they realise the important consideration of producing light and porous bread, from spoiled, or what is technically called *sour flour*. This salt which becomes wholly converted into a gaseous state during the operation of baking, causes the dough to swell up into air bubbles, which carry before them the stiff dough, and thus it renders the dough porous; the salt itself is, at the same time, totally volatilised during the operation of baking. Thus not a vestige of carbonate of ammonia remains in the bread. This salt is also largely employed by the biscuit and ginger-bread bakers.

Potatoes are likewise largely, and perhaps constantly, used by fraudulent bakers, as a cheap ingredient, to enhance their profit. The potatoes being boiled, are triturated, passed through a sieve, and incorporated with the dough by kneading. This adulteration does not materially injure the bread. The bakers assert, that the bad quality of the flour renders the addition of potatoes advantageous as well to the baker as to the purchaser, and that without this admixture in the manufacture of bread, it would be impossible to carry on the trade of a baker. But the grievance is, that the same price is taken for a potatoe loaf, as for a loaf of genuine bread, though it must cost the baker less.

I have witness, that five bushels of flour, three ounces of alum, six pounds of salt, one bushel of potatoes boiled into a stiff paste, and three quarts of yeast, with the requisite quantity of water, produce a white, light, and highly palatable bread.

Such are the artifices practised in the preparation of bread, and it must be allowed, on contrasting them with those sophistications practised by manufacturers of other articles of food, that they are comparatively unimportant. However, some medical men have no hesitation in attributing many diseases incidental to children to the use of eating adulterated bread; others again will not admit these allegations: they persuade themselves that the small quantity of alum added to the bread (perhaps upon an average, from eight to ten grains to a quartern loaf,) is absolutely harmless.

Dr. Edmund Davy, Professor of Chemistry, at the Cork Institution, has communicated the following important facts to the public concerning the manufacture of bread.

"The carbonate of magnesia of the shops, when well mixed with flour, in the proportion of from twenty to forty grains to a pound of flour, materially improves it for the purpose of making bread.

"Loaves made with the addition of carbonate of magnesia, rise well in the oven; and after being baked, the bread is light and spongy, has a good taste, and keeps well. In cases when the new flour is of an indifferent quality, from twenty to thirty grains of carbonate of magnesia to a pound of the flour will considerably improve the bread. When the flour is of the worst quality, forty grains to a pound of flour seem necessary to produce the same effect.

"As the improvement in the bread from new flour depends upon the carbonate of magnesia, it is necessary that care should be taken to mix it intimately with the flour, previous to the making of the dough.

"Mr. Davy made a great number of comparative experiments with other substances, mixed in different proportions with new bread flour. The fixed alkalies, both in their pure and carbonated state, when used in small quantity, to a certain extent were found to improve the bread made from new flour; but no substance was so efficacious in this respect as carbonate of magnesia.

"The greater number of his experiments were performed on the worst new *seconds* flour Mr. Davy could procure. He also made some trials on *seconds* and *firsts* of different quality. In some cases the results were more striking and satisfactory than in others; but in every instance the improvement of the bread, by carbonate of magnesia, was obvious.

"Mr. Davy observes, that a pound of carbonate of magnesia would be sufficient to mix with two hundred and fifty-six pounds of new flour, or at the rate of thirty grains to the pound. And supposing a pound of carbonate of magnesia to cost half-a-crown, the additional expense would be only half a farthing in the pound of flour.

"Mr. Davy conceives that not the slightest danger can be apprehended from the use of such an innocent substance, as the carbonate of magnesia, in such small proportion as is necessary to improve bread from new flour."

METHOD OF DETECTING THE PRESENCE OF ALUM IN BREAD.

Pour upon two ounces of the suspected bread, half a pint of boiling distilled water; boil the mixture for a few minutes, and filter it through unsized paper. Evaporate the fluid, to about one fourth of its original bulk, and let gradually fall into the clear fluid a solution of muriate of barytes. If a *copious* white precipitate ensues, which does not disappear by the addition of *pure*nitric acid, the presence of alum may be suspected. Bread, made without alum, produces, when assayed in this manner, merely a very slight precipitate, which originates from a minute portion of sulphate of magnesia contained in all common salt of commerce; and bread made with salt freed from sulphate of magnesia, produces an infusion with water, which does not become disturbed by the barytic test.

Other means of detecting all the constituent parts of alum, namely, the alumine, sulphuric acid, and potash, so as to render the presence of the alum unequivocal, will readily suggest itself to those who are familiar with analytical chemistry; namely: one of the readiest means is, to decompose the vegetable matter of the bread, by the action of chlorate of potash, in a platina crucible, at a red heat, and then to assay the residuary mass—by means of muriate of barytes, for sulphuric acid; by ammonia, for alumine; and by muriate of platina, for potash. The above method of detecting the presence of alum, must therefore be taken with some limitation.

There is no unequivocal test for detecting in a *ready manner* the presence of alum in bread, on account of the impurity of the common salt used in the making of bread. If we could, in the ordinary way of bread making, employ common salt, absolutely free from foreign saline substances, the mode of detecting the presence of alum, or at least one of its constituent parts, namely, the sulphuric acid, would be very easy. Some conjecture may, nevertheless, be formed of the presence, or absence, of alum, by assaying the infusion of bread in the manner stated and comparing the assay with the results afforded by an infusion of home-made or household bread, known to be genuine, and actually assayed in a similar manner.

EASY METHOD OF JUDGING OF THE GOODNESS OF BREAD CORN, AND BREAD-FLOUR.

Millers judge of the goodness of bread corn by the quantity of bran which the grain produces.

Such grains as are full and plump, that have a bright and shining appearance, without any shrivelling and shrinking in the covering of the skin, are the best; for wrinkled grains have a greater quantity of skin, or bran, than such as are sound or plump.

Pastry-cooks and bakers judge of the goodness of flour in the manner in which it comports itself in kneading. The best kind of wheaten flour assumes, at the instant it is formed into paste by the addition of water, a very gluey, ductile, and elastic paste, easy to be kneaded, and which may be elongated, flattened, and drawn in every direction, without breaking.

For the following fact we are indebted to Mr. Hatchet.

"Grain which has been heated or burnt in the stack, may in the following manner be rendered fit for being made into bread:

"The wheat must be put into a vessel capable of holding at least three times the quantity, and the vessel filled with boiling water; the grain should then be occasionally stirred, and the hollow decayed grains, which float, may be removed. When the water has become cold, or in about half an hour, it is drawn off. Then rince the corn with cold water, and, having completely drained it, spread it thinly on the floor of a kiln, and thus thoroughly dry it, stirring and turning it frequently during this part of the process."

Adulteration of Beer.

Malt liquors, and particularly porter, the favourite beverage of the inhabitants of London, and of other large towns, is amongst those articles, in the manufacture of which the greatest frauds are frequently committed.

The statute prohibits the brewer from using any ingredients in his brewings, except malt and hops; but it too often happens that those who suppose they are drinking a nutritious beverage, made of these ingredients only, are entirely deceived. The beverage may, in fact, be neither more nor less than a compound of the most deleterious substances; and it is also clear that all ranks of society are alike exposed to the nefarious fraud. The proofs of this statement will be shewn hereafter.

The author of a Practical Treatise on Brewing, which has run through eleven editions, after having stated the various ingredients for brewing porter, observes, "that however much they may surprise, however pernicious or disagreeable they may appear, he has always found them requisite in the brewing of porter, and he thinks they must invariably be used by those who wish to continue the taste, flavour, and appearance of the beer. And though several Acts of Parliament have been passed to prevent porter brewers from using many of them, yet the author can affirm, from experience, he could never produce the present flavoured porter without them. The intoxicating qualities of porter are to be ascribed to the various drugs intermixed with it. It is evident some porter is more heady than other, and it arises from the greater or less quantity of stupifying ingredients. Malt, to produce intoxication, must be used in such large quantities as would very much diminish, if not totally exclude, the brewer's profit."

The practice of adulterating beer appears to be of early date. By an Act so long ago as Queen Anne, the brewers are prohibited from mixing *cocculus indicus*, or any unwholesome ingredients, in their beer, under severe penalties: but few instances of convictions under this act are to be met with in the public records for nearly a century. To shew that they have augmented in our own days, we shall exhibit an abstract from documents laid lately before Parliament.

These will not only amply prove, that unwholesome ingredients are used by fraudulent brewers, and that very deleterious substances are also vended both to brewers and publicans for adulterating beer, but that the ingredients mixed up in the brewer's enchanting cauldron are placed above all competition, even with the potent charms of Macbeth's witches:

"Root of hemlock, digg'd i' the dark, + + + + + + + + + +For a charm of pow'rful trouble,Like a hell-broth boil and bubble;Double, double, toil and trouble,Fire burn, and cauldron bubble."

The fraud of imparting to porter and ale an intoxicating quality by narcotic substances, appears to have flourished during the period of the late French war; for, if we examine the importation lists of drugs, it will be noticed that the quantities of cocculus indicus imported in a given time prior to that period, will bear no comparison with the quantity imported in the same space of time during the war, although an additional duty was laid upon this commodity. Such has been the amount brought into this country in five years, that it far exceeds the quantity imported during twelve years anterior to the above epoch. The price of this drug has risen within these ten years from two shillings to seven shillings the pound.

It was at the period to which we have alluded, that the preparation of an extract of cocculus indicus first appeared, as a new saleable commodity, in the price-currents of *brewers'-druggists*. It was at the same time, also, that a Mr. Jackson, of notorious memory, fell upon the idea of brewing beer from various drugs, without any malt and hops. This chemist did not turn brewer himself; but he struck out the more profitable trade of teaching his mystery to the brewers for a handsome fee. From that time forwards, written directions, and recipe-books for using the chemical preparations to be substituted for malt and hops, were respectively sold; and many adepts soon afterwards appeared every where, to instruct brewers in the nefarious practice, first pointed out by Mr. Jackson. From that time, also, the fraternity of brewers'-chemists took its rise. They made it their chief business to send travellers all over the country with lists and samples exhibiting the price and quality of the articles manufactured by them for the use of brewers only. Their trade spread far and wide, but it was amongst the country brewers chiefly that they found the most customers; and it is amongst them, up to the present day, as I am assured by some of these operators, on whose veracity I can rely, that the greatest quantities of unlawful ingredients are sold.

The Act of Parliament prohibits chemists, grocers, and druggists, from supplying illegal ingredients to brewers under a heavy penalty, as is obvious from the following abstract of the Act.

"No druggist, vender of, or dealer in drugs, or chemist, or other person, shall sell or deliver to any licensed brewer, dealer in or retailer of beer, knowing him to be such, or shall sell or deliver to any person on account of or in trust for any such brewer, dealer or retailer, any liquor called by the name of or sold as colouring, from whatever material the same may be made, or any material or preparation other than unground brown malt for darkening the colour of worts or beer, or any liquor or preparation made use of for darkening the colour of worts or beer, or any molasses, honey, vitriol, quassia, cocculus Indian, grains of paradise, Guinea pepper or opium, or any extract or preparation of molasses, or any article or preparation to be used in worts or beer for or as a substitute for malt or hops; and if any druggist shall offend in any of these particulars, such liquor preparation, molasses, &c. shall be forfeited, and may be seized by any officer of excise, and the person so offending shall for each offence forfeit 500*l*."

The following is a list of druggists and grocers, prosecuted by the Court of Excise, and convicted of supplying unlawful ingredients to brewers.

List of Druggists and Grocers, prosecuted and convicted from 1812 to 1819, for supplying illegal Ingredients to Brewers for adulterating Beer.

John Dunn and another, druggists, for selling adulterating ingredients to brewers, verdict 500*l*.

George Rugg and others, druggists, for selling adulterating ingredients to brewers, verdict 500*l*.

John Hodgkinson and others, for selling adulterating ingredients to brewers, 100*l.* and costs.

William Hiscocks and others, for selling adulterating ingredients to a brewer, 200*l.* and costs.

G. Hornby; for selling adulterating ingredients to a brewer, 200*l.*

W. Wilson, for selling adulterating ingredients to a brewer, 200*l.*

George Andrews, grocer, for selling adulterating ingredients to a brewer, 25*l.* and costs.

Guy Knowles, for selling substitute for hops, costs.

Kernot and Alsop, for selling cocculus india, &c. 25*l.*

Joseph Moss, for selling various drugs, 300*l.*

Ph. Whitcombe, John Dunn, and Arthur Waller, druggists, for having liquor for darkening the colour of beer, hid and concealed.

Isaac Hebberd, for having liquor for darkening the colour of beer, hid and concealed.

Ph. Whitcombe, John Dunn, and Arthur Waller, druggists, for making liquor for darkening the colour of beer.

John Lord, grocer, for selling molasses to a brewer, 20*l.* and costs.

John Smith Carr, grocer, for selling molasses to a brewer, 20*l.* and costs.

Edward Fox, grocer, for selling molasses to a brewer, 25*l.* and costs.

John Cooper, grocer, for selling molasses to a brewer, 40*l.* and costs.

Joseph Bickering, grocer, for selling molasses to a brewer, 40*l.* and costs.

John Howard, grocer, for selling molasses to a brewer, 25*l.* and costs.

James Reynolds, grocer, for selling molasses to a brewer, costs.

Thomas Hammond, grocer, for selling molasses to a brewer, 20*l.* and costs.

J. Mackway, grocer, for selling molasses to a brewer, 20*l.*

T. Renton, grocer, for selling molasses to a brewer, costs, and taking out a license.

R. Adamson, grocer, for selling molasses to a brewer, costs, and taking out a license.

W. Weaver, for selling Spanish liquorice to a brewer, 200*l.*

J. Moss, for selling Spanish liquorice to a brewer.

Alex. Braden, for selling liquorice, 20*l.*

J. Draper, for selling molasses to a brewer, 20*l.*

PORTER.

The method of brewing porter has not been the same at all times as it is at present.

At first, the only essential difference in the methods of brewing this liquor and that of other kinds of beer, was, that porter was brewed from brown malt only; and this gave to it both the colour and flavour required. Of late years it has been brewed from mixtures of pale and brown malt.

These, at some establishments, are mashed separately, and the worts from each are afterwards mixed together. The proportion of pale and brown malt, used for brewing porter, varies in different breweries; some employ nearly two parts of pale malt and one part of brown malt; but each brewer appears to have his own proportion; which the intelligent manufacturer varies, according to the nature and qualities of the malt. Three pounds of hops are, upon an average, allowed to every barrel, (thirty-six gallons) of porter.

When the price of malt, on account of the great increase in the price of barley during the late war, was very high, the London brewers discovered that a larger quantity of wort of a given strength could be obtained from pale malt than from brown malt. They therefore increased the quantity of the former and diminished that of the latter. This produced beer of a paler colour, and of a less bitter flavour. To remedy these disadvantages, they invented an artificial colouring substance, prepared by boiling brown sugar till it acquired a very dark brown colour; a solution of which was employed to darken the colour of the beer. Some brewers made use of the infusion of malt instead of sugar colouring. To impart to the beer a bitter taste, the fraudulent brewer employed quassia wood and wormwood as a substitute for hops.

But as the colouring of beer by means of sugar became in many instances a pretext for using illegal ingredients, the Legislature, apprehensive from the mischief that might, and actually did, result from it, passed an Act prohibiting the use of burnt sugar, in July 1817; and nothing but malt and hops is now allowed to enter into the composition of beer: even the use of isinglass for clarifying beer, is contrary to law.

No sooner had the beer-colouring Act been repealed, than other persons obtained a patent for effecting the purpose of imparting an artificial colour to porter, by means of brown malt, specifically prepared for that purpose only. The beer, coloured by the new method, is more liable to become spoiled, than when coloured by the process formerly practised. The colouring malt does not contain any considerable portion of saccharine matter. The grain is by mere torrefaction converted into a gum-like substance, wholly soluble in water, which renders the beer more liable to pass into the acetous fermentation than the common brown malt is capable of doing; because the latter, if prepared from good barley, contains a portion of saccharine matter, of which the patent malt is destitute.

But as brown malt is generally prepared from the worst kind of barley, and as the patent malt can only be made from good grain, it may become, on that account, an useful article to the brewer (at least, it gives colour and body to the beer;) but it cannot materially economise the quantity of malt necessary to produce good porter. Some brewers of eminence in this town have assured me, that the use of this mode of colouring beer is wholly unnecessary; and that porter of the requisite colour may be brewed better without it; hence this kind of malt is not used in their establishments. The quantity of gum-like matter which it contains, gives too much ferment to the beer, and renders it liable to spoil. Repeated experiments, made on a large scale, have settled this fact.

STRENGTH AND SPECIFIC DIFFERENCES OF DIFFERENT KINDS OF PORTER.

The strength of all kinds of beer, like that of wine, depends on the quantity of spirit contained in a given bulk of the liquor.

The reader need scarcely be told, that of no article there are more varieties than of porter. This, no doubt, arises from the different mode of manufacturing the beer, although the ingredients are the same. This difference is more striking in the porter manufactured among country brewers, than it is in the beer brewed by the eminent London porter brewers. The totality of the London porter exhibits but very slight differences, both with respect to strength or quantity of spirit, and solid extractive matter, contained in a given bulk of it. The spirit may be stated, upon an average, to be 4,50 per cent. in porter retailed at the publicans; the solid matter, is from twenty-one to twenty-three pounds per barrel of thirty-six gallons. The country-brewed porter is seldom well fermented, and seldom contains so large a quantity of spirit; it usually abounds in mucilage; hence it becomes turbid when mixed with alcohol. Such beer cannot keep, without becoming sour.

It has been matter of frequent complaint, that all the porter now brewed, is not what porter was formerly. This idea may be true with some exceptions. My professional occupations have, during these twenty-eight years, repeatedly obliged me to examine the strength of London porter, brewed by different brewers; and, from the minutes made on that subject, I am authorised to state, that the porter now brewed by the eminent London brewers, is unquestionably stronger than that which was brewed at different periods during the late French war. Samples of brown stout with which I have been obligingly favoured, whilst writing this Treatise, by Messrs. Barclay, Perkins, and Co.—Messrs. Truman, Hanbury, and Co.—Messrs. Henry Meux and Co.—and other eminent brewers of this capital—afforded, upon an average, 7,25 per cent. of alcohol, of 0,833 specific gravity; and porter, from the same houses, yielded upon an average 5,25 per cent. of alcohol, of the same specific gravity; this beer received from the brewers was taken from the same store from which the publicans are supplied.

It is nevertheless singular to observe, that from fifteen samples of beer of the same denominations, procured from different retailers, the proportions of spirit fell considerably short of the above quantities. Samples of brown stout, procured from the retailers, afforded, upon an average, 6,50 per cent. of alcohol; and the average strength of the porter was 4,50 per cent. Whence can this difference between the beer furnished by the brewer, and that retailed by the publican, arise? We shall not be at a loss to answer this question, when we find that so many retailers of porter have been prosecuted and convicted for mixing table beer with their strong beer; this is prohibited by law, as becomes obvious by the following words of the Act.

"If any common or other brewer, innkeeper, victualler, or retailer of beer or ale, shall mix or suffer to be mixed any strong beer, ale, or worts, with table beer, worts, or water, in any tub or measure, he shall forfeit 50*l*." The difference between strong and table beer, is thus settled by Parliament.

"All beer or ale above the price of eighteen shillings per barrel, exclusive of ale duties now payable (viz. ten shillings per barrel,) or that may be hereafter payable in respect thereof, shall be deemed strong beer or ale; and all beer of the price of eighteen shillings the barrel or under, exclusive of the duty payable (viz. two shillings per barrel) in respect thereof, shall be deemed table beer within the meaning of this and all other Acts now in force, or that may hereafter be passed in relation to beer or ale or any duties thereon."

List of Publicans prosecuted and convicted from 1815 to 1818, for adulterating Beer with illegal Ingredients, and for mixing Table Beer with their Strong Beer.

William Atterbury, for using salt of steel, salt, molasses, &c. and for mixing table beer with strong beer, 40*l.*

Richard Dean, for using salt of steel, salt, molasses, &c. and for mixing table beer with strong beer, 50*l.*

John Jay, for using salt of steel, salt, molasses, &c. and for mixing table beer with strong beer, 50*l.*

James Atkinson, for using salt of steel, salt, molasses, &c. and for mixing table beer with strong beer, 20*l.*

Samuel Langworth, for using salt of steel, salt, molasses, &c. and for mixing table beer with strong beer, 50*l.*

Hannah Spencer, for using salt of steel, salt, molasses, &c. and for mixing table beer with strong beer, 150*l.*

— — Hoeg, for using salt of steel, salt, molasses, &c. and for mixing table beer with strong beer, 5*l.*

Richard Craddock, for using salt of steel, salt, molasses, &c. and for mixing table beer with strong beer, 100*l.*

James Harris, for using salt of steel, salt, molasses, &c. and for receiving stale beer, and mixing it with strong beer, 42*l.* and costs.

Thomas Scoons, for using salt of steel, salt, molasses, &c. and for mixing stale beer with strong beer, verdict 200*l.*

Diones Geer and another, for using salt of steel, salt, molasses, &c. and for mixing strong and table beer, verdict 400*l.*

Charles Coleman, for using salt of steel, salt, molasses, &c. and for mixing strong and table beer, 35*l.* and costs.

William Orr, for using salt of steel, salt, molasses, &c. and for mixing strong and table beer, 50*l.*

John Gardiner, for using salt of steel, salt, molasses, &c. and for mixing strong and table beer, 100*l.*

John Morris, for using salt of steel, salt, molasses, &c. and for mixing strong and table beer, 20*l.*

John Harbur. for using salt of steel, salt, molasses, &c. and for mixing strong and table beer, 50*l.*

John Corrie, for mixing strong beer with table beer.

John Cape, for mixing strong beer with table beer.

Joseph Gudge, for mixing strong beer with small beer.

ILLEGAL SUBSTANCES USED FOR ADULTERATING BEER.

We have stated already that nothing is allowed by law to enter into the composition of beer, but malt and hops.

The substances used by fraudulent brewers for adulterating beer, are chiefly the following:

Quassia, which gives to beer a bitter taste, is substituted for hops; but hops possesses a more agreeable aromatic flavour, and there is also reason to believe that they render beer less liable to spoil by keeping; a property which does not belong to quassia. It requires but little discrimination to distinguish very clearly the peculiar bitterness of quassia in adulterated porter. Vast quantities of the shavings of this wood are sold in a half-torrefied and ground state to disguise its obvious character, and to prevent its being recognised among the waste materials of the brewers. Wormwood has likewise been used by fraudulent brewers.

The adulterating of hops is prohibited by the Legislature.

"If any person shall put any drug or ingredient whatever into hops to alter the colour or scent thereof, every person so offending, convicted by the oath of one witness before one justice of peace for the county or place where the offence was committed, shall forfeit 5*l.* for every hundred weight."

Beer rendered bitter by quassia never keeps well, unless it be kept in a place possessing a temperature considerably lower than the temperature of the surrounding atmosphere; and this is not well practicable in large establishments.

The use of boiling the wort of beer with hops, is partly to communicate a peculiar aromatic flavour which the hop contains, partly to cover the sweetness of undecomposed saccharine matter, and also to separate, by virtue of the gallic acid and tannin it contains, a portion of a peculiar vegetable mucilage somewhat resembling gluten, which is still diffused through the beer. The compound thus produced, separates in small flakes like those of curdled soap; and by these means the beer is rendered less liable to spoil. For nothing contributes more to the conversion of beer, or any other vinous fluid, into vinegar, than mucilage. Hence, also, all full-bodied and clammy ales, abounding in mucilage, and which are generally ill fermented, do not keep as perfect ale ought to do. Quassia is, therefore, unfit as a substitute for hops; and even English hops are preferable to those imported from the Continent; for nitrate of silver and acetate of lead produce a more abundant precipitate from an infusion of English hops, than can be obtained from a like infusion by the same agents from foreign hops.

One of the qualities of good porter, is, that it should bear *a fine frothy head*, as it is technically termed: because professed judges of this beverage, would not pronounce the liquor excellent, although it possessed all other good qualities of porter, without this requisite.

To impart to porter this property of frothing when poured from one vessel into another, or to produce what is also termed a *cauliflower head*, the mixture called *beer-heading*, composed of common green vitriol (sulphate of iron,) alum, and salt, is added. This addition to the beer is generally made by the publicans. It is unnecessary to genuine beer, which of itself possesses the property of bearing a strong white froth, without these additions; and it is only in consequence of table beer being mixed with strong beer, that the frothing property of the porter is lost. From experiments I have tried on this subject, I have reason to believe that the sulphate of iron, added for that purpose, does not possess the power ascribed to it. But the publicans frequently, when they fine a butt of beer, by means of isinglass, adulterate the porter at the same time with table beer, together with a quantity of molasses and a small portion of extract of gentian root, to keep up the peculiar flavour of the porter; and it is to the molasses chiefly, which gives a spissitude to the beer, that the frothing property must be ascribed; for, without it, the sulphate of iron does not produce the property of frothing in diluted beer.

Capsicum and grains of paradise, two highly acrid substances, are employed to give a pungent taste to weak insipid beer. Of late, a concentrated tincture of these articles, to be used for a similar purpose, and possessing a powerful effect, has appeared in the price-currents of brewers' druggists. Ginger root, coriander seed, and orange peels, are employed as flavouring substances chiefly by the ale brewers.

From these statements, and the seizures that have been made of illegal ingredients at various breweries, it is obvious that the adulterations of beer are not imaginary. It will be noticed, however, that some of the sophistications are comparatively harmless, whilst others are effected by substances deleterious to health.

The following list exhibits some of the unlawful substances seized at different breweries and at chemical laboratories.

List of Illegal Ingredients, seized from 1812 to 1818, at various Breweries and Brewers' Druggists.

1812, July. Josiah Nibbs, at Tooting, Surrey.

Multum	84	lbs.
Cocculus indicus	12	
Colouring	4	galls.
Honey	about 180 lbs.	
Hartshorn Shavings	14	
Spanish Juice	46	
Orange Powder	17	
Ginger	56	

Penalty 300*l.*

1813, June 13. Sarah Willis, at West Ham, Essex.

Cocculus indicus	1	lb.
Spanish Juice	12	
Hartshorn Shavings	6	
Orange Powder	1	

Penalty 200*l.*

August 3. Cratcherode Whiffing, Limehouse.

Grains of Paradise	44	lbs.
Quassia	10	
Liquorice	64	
Ginger	80	
Caraway Seeds	40	
Orange Powder	14	
Copperas	4	

Penalty 200*l.*

Nov. 25. Elizabeth Hasler, at Stratford.

Cocculus indicus	12	lbs.
Multum	26	
Grains of Paradise	12	
Spanish Juice	30	
Orange Powder	3	

Penalty 200*l.*

Dec. 14. John Abbott, at Canterbury, Kent.

Copperas, &c.	14	lbs.
Orange powder	2	

Penalty 500*l.*, and Crown's costs.

Proof of using drugs at various times.

1815, Feb. 15. Mantel and Cook, Castle-street, Bloomsbury-square.

Proof of mixing strong with table beer, and using colouring and other things.

Compromised for 300*l.*

1817. From Peter Stevenson, an old Servant to Dunn and Waller, St. John-street, brewers' druggists.

Cocculus Indicus Extract	6	lbs.
Multum	560	
Capsicum	88	
Copperas	310	
Quassia	150	
Colouring and Drugs	84	
Mixed Drugs	240	
Spanish Liquorice	420	
Hartshorn Shavings	77	
Liquorice Powder	175	
Orange powder	126	
Caraway Seeds	100	
Ginger	110	
Ginger Root	176	

Condemned, not being claimed.

July 30. Luke Lyons, Shadwell.

Capsicum	1	lb
Liquorice Root Powder	2	
Coriander Seed	2	
Copperas	1	
Orange Powder	8	
Spanish Liquorice	1/2	
Beer Colouring	24	galls

Not tried. (7th May, 1818.)

Aug. 6. John Gray, at West Ham.

Multum	4	lbs.
Spanish Liquorice	21	
Liquorice Root Powder	113	
Ginger	116	
Honey	11	

> Penalty, 300*l.*, and costs; including mixing strong beer with table, and paying table-beer duty for strong beer, &c.

Numerous other seizures of illegal substances, made at breweries, might be advanced, were it necessary to enlarge this subject to a greater extent.

Mr. James West, from the excise office, being asked in the Committee of the House of Commons, appointed, 1819, to examine and report on the petition of several inhabitants of London, complaining of the high price and inferior quality of beer, produced the following seized articles:—"One bladder of honey, one bladder of extract of cocculus indicus, ground guinea pepper or capsicum, vitriol or copperas, orange powder, quassia, ground beer-heading, hard multum, another kind of multum or beer preparation, liquorice powder, and ground grains of paradise."

Witness being asked "Where did you seize these things?" Answer, "Some of them were seized from brewers, and some of them from brewers' druggists, within these two years past." (May 8, 1818.)

Another fraud frequently committed, both by brewers and publicans, (as is evident from the Excise Report,) is the practice of adulterating strong beer with small beer—This fraud is prohibited by law, since both the revenue and the public suffer by it. "The duty upon strong beer is ten shillings a barrel; and upon table beer it is two shillings. The revenue suffers, because a larger quantity of beer is sold as strong beer; that is, at a price exceeding the price of table beer, without the strong beer duty being paid. In the next place, the brewer suffers, because the retailer gets table or mild beer, and retails it as strong beer." The following are the words of the Act, prohibiting the brewers mixing table beer with strong beer.

"If any common brewer shall mix or suffer to be mixed any strong beer, or strong worts with table beer or table worts, or with water in any guile or fermenting tun after the declaration of the quantity of such guile shall have been made; or if he shall at any time mix or suffer to be mixed strong beer or strong worts with table beer worts or with water, in any vat, cask, tub, measures or utensil, not being an entered guile or fermenting tun, he shall forfeit 200 pounds."

With respect to the persons who commit this offence, Mr. Carr, the Solicitor of the Excise, observes, that "they are generally brewers who carry on the double trade of brewing both strong and table beer. It is almost impossible to prevent them from mixing one with the other; and frauds of very great extent have been detected, and the parties punished for that offence. One brewer at Plymouth evaded duties to the amount of 32,000 pounds; and other brewers, who brew party guiles of beer, carrying on the two trades of ale and table beer brewers, where the trade is a victualling brewer, which is different from the common brewer, he being a person who sells only wholesale; the victualling brewer being a brewer and also a seller by retail."

"In the neighbourhood of London," Mr. Carr continues, "more particularly, I speak from having had great experience, from the informations and evidence which I have received, that the retailers carry on a most extensive fraud upon the public, in purchasing stale table beer, or the bottoms of casks. There are a class of men who go about and sell such beer at table-beer price to public victuallers, who mix it in their cellars. If they receive beer from their brewers which is mild, they purchase stale beer; and if they receive stale beer, they purchase common table beer for that purpose; and many of the prosecutions are against retailers for that offence." The following may serve in proof of this statement.

List of Brewers prosecuted and convicted from 1813 to 1819, for adulterating Strong Beer with Table Beer.

Thomas Manton and another, brewers, for mixing strong and table beer, verdict 300*l*.

Mark Morrell and another, brewers, for mixing strong and table beer, 20*l*. and costs.

Robert Jones and another, brewers, for mixing strong and table beer, verdict 125*l*.

Robert Stroad, brewer, for mixing strong and table beer, 200*l.* and costs.

William Cobbett, brewer, mixing strong and table beer, 100*l.* and costs.

Thomas Richard Withers, brewer, for mixing strong and table beer, 75*l.* and costs.

John Cowel, brewer, for mixing table beer with strong, 50*l.* and costs.

John Mitchell, brewer, for mixing table beer with strong, absconded.

George Lloyd and another, brewers, for mixing table beer with strong, 25*l.* and costs.

James Edmunds and another, brewers, for mixing table beer with strong, for a long period, verdict 600*l.*

John Hoffman, brewer, for mixing strong and table beer, and using molasses, 130*l.* and costs.

Samuel Langworth, brewer, for mixing strong with stale table beer, 10*l.* and costs.

Hannah Spencer, brewer, for mixing strong with stale table beer, verdict 150*l.*

Joseph Smith and others, brewers, for mixing strong and table beer.

Philip George, brewer, for mixing strong and table beer, verdict 200*l.*

Joshua Row, brewer, for mixing strong and table beer, verdict 400*l.*

John Drew, jun. and another, for mixing strong beer with table, 50*l.* and costs.

John Cape, brewer, for mixing strong and table beer, 250*l.* and costs.

John Williams and another, brewers, for mixing strong and table beer, verdict 200*l.*

OLD, OR ENTIRE; AND NEW, OR MILD BEER.

It is necessary to state, that every publican has two sorts of beer sent to him from the brewer; the one is called *mild*, which is beer sent out fresh as it is brewed; the other is called *old*; that is, such as is brewed on purpose for keeping, and which has been kept in store a twelve-month or eighteen months. The origin of the beer called *entire*, is thus related by the editor of the Picture of London: "Before the year 1730, the malt liquors in general used in London were ale, beer, and two-penny; and it was customary to call for a pint, or tankard, of half-and-half, *i.e.*half of ale and half of beer, half of ale and half of two-penny. In course of time it also became the practice to call for a pint or tankard of *three-threads*, meaning a third of ale, beer, and two-penny; and thus the publican had the trouble to go to three casks, and turn three cocks, for a pint of liquor. To avoid this inconvenience and waste, a brewer of the name of Harwood conceived the idea of making a liquor, which should partake of the same united flavours of ale, beer, and two-penny; he did so, and succeeded, calling it *entire*, or entire butt, meaning that it was drawn entirely from one cask or butt; and as it was a very hearty and nourishing liquor, and supposed to be very suitable for porters and other working people, it obtained the name of*porter*." The system is now altered, and porter is very generally compounded of two kinds, or rather the same liquor in two different states, the due admixture of which is palatable, though neither is good alone. One is *mild* porter, and the other *stale* porter; the former is that which has a slightly bitter flavour; the latter has been kept longer. This mixture the publican adapts to the palates of his several customers, and effects the mixture very readily, by means of a machine, containing small pumps worked by handles. In these are four pumps, but only three spouts, because two of the pumps throw out at the same spout: one of these two pumps draws the mild, and the other the stale porter, from the casks down in the cellar; and the publican, by dexterously changing his hold works either pump, and draws both kinds of beer at the same spout. An indifferent observer supposes, that since it all comes from one spout, it is entire butt beer, as the publican professes over his door, and which has been decided by vulgar prejudice to be only good porter, though the difference is not easily distinguished. I have been informed by several eminent brewers, that of late, a far greater quantity is consumed of mild than of stale beer.

The entire beer of the modern brewer, according to the statement of C. Barclay, Esq. "consists of some beer brewed expressly for the purpose of keeping: it likewise contains a portion of returns from publicans; a portion of beer from the bottoms of vats; the beer that is drawn off from the pipes, which convey the beer from one vat to another, and from one part of the premises to another. This beer is collected and put into vats. Mr. Barclay also states that it contains a certain portion of brown stout, which is twenty shillings a barrel dearer than common beer; and some bottling beer, which is ten shillings a barrel dearer; and that all these beers, united, are put into vats, and that it depends upon various circumstances, how long they may remain in those vats before they become perfectly bright. When bright, this beer is sent out to the publicans, for their *entire* beer, and there is sometimes a small quantity of mild beer mixed with it."

The present entire beer, therefore, is a very heterogeneous mixture, composed of all the waste and spoiled beer of the publicans—the bottoms of butts—the leavings of the pots—the drippings of the machines for drawing the beer—the remnants of beer that lay in the leaden pipes of the brewery, with a portion of brown stout, bottling beer, and mild beer.

The old or *entire* beer we have examined, as obtained from Messrs. Barclay's, and other eminent London brewers, is unquestionably a good compound; but it does no longer appear to be necessary, among fraudulent brewers, to brew beer on purpose for keeping, or to keep it twelve or eighteen months. A more easy, expeditious, and economical method has been discovered to convert any sort of beer into entire beer, merely by the admixture of a portion of sulphuric acid. An imitation of the age of eighteen months is thus produced in an instant. This process is technically called to bring beer *forward*, or to make it *hard*.

The practice is a bad one. The genuine, old, or entire beer, of the honest brewer, is quite a different compound; it has a rich, generous, full-bodied taste, without being acid, and a vinous odour: but it may, perhaps, not be generally known that this kind of beer always affords a less proportion of alcohol than is produced from mild beer. The practice of bringing beer *forward,* it is to be understood, is resorted to only by fraudulent brewers.

If, on the contrary, the brewer has too large a stock of old beer on his hands, recourse is had to an opposite practice of converting stale, half-spoiled, or sour beer, into mild beer, by the simple admixture of an alkali, or an alkaline earth. Oyster-shell powder and subcarbonate of potash, or soda, are usually employed for that purpose. These substances neutralise the excess of acid, and render sour beer somewhat palatable. By this process the beer becomes very liable to spoil.

It is the worst expedient that the brewer can practise: the beer thus rendered *mild,* soon loses its vincus taste; it becomes vapid; and speedily assumes a muddy grey colour, and an exceedingly disagreeable taste.

These sophistications may be considered, at first, as minor crimes practised by fraudulent brewers, when compared with the methods employed by them for rendering beer noxious to health by substances absolutely injurious.

To increase the intoxicating quality of beer, the deleterious vegetable substance, called *cocculus indicus,* and the extract of this poisonous berry, technically called *black extract*, or, by some,*hard multum*, are employed. Opium, tobacco, nux vomica, and extract of poppies, have also been used.

This fraud constitutes by far the most censurable offence committed by unprincipled brewers; and it is a lamentable reflection to behold so great a number of brewers prosecuted and convicted of this crime; nor is it less deplorable to find the names of druggists, eminent in trade, implicated in the fraud, by selling the unlawful ingredients to brewers for fraudulent purposes.

List of Brewers prosecuted and convicted from 1813 to 1819, for receiving and using illegal Ingredients in their Brewings.

Richard Gardner, brewer, for using adulterating ingredients, 100*l.*, judgment by default.

Stephen Webb and another, brewers, for using adulterating ingredients, and mixing strong and table beer, verdict 500*l.*

Henry Wyatt, brewer, for using adulterating ingredients, verdict 400*l.*

John Harbart, retailer, for receiving adulterating ingredients, verdict 150*l.*

Philip Blake and others, brewers, for using adulterating ingredients, and mixing strong and table beer, verdict 250*l.*

James Sneed, for receiving adulterating ingredients, 25*l.* and costs.

John Rewell and another, brewers, ditto, verdict 100*l.*

John Swain and another, ditto, for using adulterating ingredients, verdict 200*l.*

John Ing, brewer, ditto, stayed on defendant's death.

John Hall, ditto, for receiving adulterating ingredients, 5*l.* and costs.

John Webb, retailer, for using adulterating ingredients.

Ralph Fogg and another, brewers, for receiving and using adulterating ingredients.

John Gray, brewer, for using adulterating ingredients, 300*l.* and costs.

Richard Bowman, for using liquid in bladder, supposed to be extract of cocculus, 100*l.*

Richard Bowman, brewer, for ditto, 100*l.* and costs.

Septimus Stephens. brewer, for ditto, verdict 50*l.*

James Rogers and another, brewer, for ditto, 220*l.* and costs.

George Moore, brewer, for using colouring, 300*l.* and costs.

John Morris, for using adulterating ingredients.

Webb and Ball, for using ginger, Guinea pepper, and brown powder, (name unknown), 1st 100*l.* 2nd 500*l.*

Henry Clarke, for using molasses, 150*l.*

Kewell and Burrows, for using cocculus india, multum, &c. 100*l.*

Allatson and Abraham, for using cocculus india, multum, and porter flavour, 630*l.*

Swain and Sewell, for using cocculus india, Guinea-opium, &c. 200*l.*

John Ing, for using cocculus india, hard colouring, and honey, *dead.*

William Dean, for using molasses, 50*l.*

John Cowell, for using Spanish-liquorice, and mixing table beer with strong beer, 50*l.*

John Mitchell, for using cocculus india, vitriol, and Guinea pepper, *left the country.*

Lloyd and Man, for using extract of cocculus, 25*l.*

John Gray, for using ginger, hartshorn shavings, and molasses, 300*l.*

Jon Hoffman, for using molasses, Spanish juice, and mixing table with strong beer, 130*l.*

Rogers and Boon, for using extract of cocculus, multum, porter flavour, &c. 220*l.*

—— Betteley, for using wormwood, coriander seed, and Spanish juice, 200*l.*

William Lane, brewer, for using wormwood instead of hops, 5*l.* and costs.

That a minute portion of an unwholesome ingredient, daily taken in beer, cannot fail to be productive of mischief, admits of no doubt; and there is reasons to believe that a small quantity of a narcotic substance (and cocculus indicus is a powerful narcotic), daily taken into the stomach, together with an intoxicating liquor, is highly more efficacious than it would be without the liquor. The effect may be gradual; and a strong constitution, especially if it be assisted with constant and hard labour, may counteract the destructive consequences perhaps for many years; but it never fails to shew its baneful effects at last. Independent of this, it is a well-established fact, that porter drinkers are very liable to apoplexy and palsy, without taking this narcotic poison.

If we judge from the preceding lists of prosecutions and convictions furnished by the Solicitor of the Excise, it will be evident that many wholesale brewers, as well as retail dealers, stand very conspicuous among those offenders. But the reader will likewise notice, that there are no convictions, in any instance, against any of the eleven great London porter brewers for any illegal practice. The great London brewers, it appears, believe that the publicans alone adulterate the beer. That many of the latter have been convicted of this fraud, the Report of the Board of Excise amply shews.—

The following statement relating to this subject, we transcribe from a Parliamentary document:

Mr. Perkins being asked, whether he believed that any of the inferior brewers adulterated beer, answered, "I am satisfied there are some instances of that."

Question.—"Do you believe publicans do?" *Answer.*—"I believe they do." *Q.*—"To a great extent?" *A.*—"Yes." *Q.*—"Do you believe they adulterate the beer you sell them?" *A.*—"I am satisfied there are some instances of that."—Mr. J. Martineau being asked the following

Question.—"In your judgment is any of the beer of the metropolis, as retailed to the publican, mixed with any deleterious ingredients?"

Answer.—"In retailing beer, in some instances, it has been."

Question.—"By whom, in your opinion, has that been done?"

Answer.—"In that case by the publicans who vend it."

On this point, it is but fair, to the minor brewers, to record also the answers of some officers of the revenue, when they were asked whether they considered it more difficult to detect nefarious practices in large breweries than in small ones.

Mr. J. Rogers being thus questioned in the Committee of the House of Commons, "Supposing the large brewers to use deleterious or any illegal ingredients to such an amount as could be of any importance to their concern, do you think it would, or would not, be more easy to detect it in those large breweries, than in small ones?" his answer was, "more difficult to detect it in the large ones:" and witness being asked to state the reason why, answered, "Their premises are so much larger, and there is so much more strength, that a cart load or two is got rid of in a minute or two." Witness "had known, in five minutes, twenty barrels of molasses got rid of as soon as the door was shut."

Another witness, W. Wells, an excise officer, in describing the contrivances used to prevent detection, stated, that at a brewer's, at Westham, the adulterating substances "were not kept on the premises, but in the brewer's house; not the principal, but the working brewers; it not being considered, when there, as liable to seizure: the brewer had a very large jacket made expressly for that purpose, with very large pockets; and, on brewing mornings, he would take his pockets full of the different ingredients. Witness supposed that such a man's jacket, similar to what he had described, would convey quite sufficient for any brewery in England, as to *cocculus indicus*."

That it may be more difficult for the officers of the excise to detect fraudulent practices in large breweries than in small ones, may be true to a certain extent: but what eminent London porter brewer would stake his reputation on the chance of so paltry a gain, in which he would inevitably be at the mercy of his own man? The eleven great porter brewers of this metropolis are persons of so high respectability, that there is no ground for the slightest suspicion that they would attempt any illegal practices, which they were aware could not possibly escape detection in their extensive establishments. And let it be remembered, that none of them have been detected for any unlawful practices, with regard to the processes of their manufacture, or the adulteration of their beer.

METHOD OF DETECTING THE ADULTERATION OF BEER.

The detection of the adulteration of beer with deleterious vegetable substances is beyond the reach of chemical analysis. The presence of sulphate of iron may be detected by evaporating the beer to perfect dryness, and burning away the vegetable matter obtained, by the action of chlorate of pot-ash in a red-hot crucible. The sulphate of iron will be left behind among the residue in the crucible, which when dissolved in water, may be assayed, for the constituent parts of the salt, namely, iron and sulphuric acid: for the former, by tincture of galls, ammonia, and prussiate of potash; and for the latter, by muriate of barytes.

Beer, which has been rendered fraudulently *hard* by the admixture of sulphuric acid, affords a white precipitate (sulphate of barytes), by dropping into it a solution of acetate or muriate of barytes; and this precipitate, when collected by filtering the mass, and after having been dried, and heated red-hot for a few minutes in a platina crucible, does not disappear by the addition of nitric, or muriatic acid. Genuine old beer may produce a precipitate; but the precipitate which it affords, after having been made red-hot in a platina crucible, instantly becomes re-dissolved with effervescence by pouring on it some pure nitric or muriatic acid; in that case the precipitate is malate (not sulphate) of barytes, and is owing to a portion of malic acid having been formed in the beer.

But with regard to the vegetable materials deleterious to health, it is extremely difficult, in any instance, to detect them by chemical agencies; and in most cases it is quite impossible, as in that of cocculus indicus in beer.

METHOD OF ASCERTAINING THE QUANTITY OF SPIRIT CONTAINED IN PORTER, ALE, OR OTHER KINDS OF MALT LIQUORS.

Take any quantity of the beer, put it into a glass retort, furnished with a receiver, and distil, with a gentle heat, as long as any spirit passes over into the receiver; which may be known by heating from time to time a small quantity of the obtained fluid in a tea-spoon over a candle, and bringing into contact with the vapour of it the flame of a piece of paper. If the vapour of the distilled fluid catches fire, the distillation must be continued until the vapour ceases to be set on fire by the contact of a flaming body. To the distilled liquid thus obtained, which is the spirit of the beer, combined with water, add, in small quantities at a time, pure subcarbonate of potash (previously freed from water by having been exposed to a red heat,) till the last portion of this salt added, remains undissolved in the fluid. The spirit will thus become separated from the water, because the subcarbonate of potash abstracts from it the whole of the water which it contained; and this combination sinks to the bottom, and the spirit alone floats on the top. If this experiment be made in a glass tube, about half or three-quarters of an inch in diameter, and graduated into 50 or 100 equal parts, the relative per centage of spirit in a given quantity of beer may be seen by mere inspection.

Quantity of Alcohol contained in Porter, Ale, and other kinds of Malt Liquors.

One hundred parts, by Measure, contained.	Parts of Alcohol, by Measure.
Ale, home-brewed	8,30
Ale, Burton, three Samples	6,25
Ale, Burton	8,88
Ale, Edinburgh	6,20
Ale, Dorchester	5,50
Ale, common London-brewed, six samples	5,82
Ale, Scotch, three samples	5,75
Porter, London, eight samples	4,00
Ditto, Ditto	4,20
Ditto, Ditto	4,45
Ditto, Ditto, bottled.	4,75
Brown Stout, four samples	5

Ditto, Ditto	6,80
Small Beer, six samples	0,75
Ditto, Ditto	1,28

Counterfeit Tea-Leaves.

The late detections that have been made respecting the illicit establishments for the manufacture of imitation tea leaves, arrested, not long ago, the attention of the public; and the parties by whom these manufactories were conducted, together with the numerous venders of the factitious tea, did not escape the hand of justice. In proof of this statement, it is only necessary to consult the London newspapers (the Times and the Courier) from March to July 1818; which show to what extent this nefarious traffic has been carried on; and they report also the prosecutions and convictions of numerous individuals who have been guilty of the fraud. The following are some of those prosecutions and convictions.

Hatton Garden.—On Saturday an information came to be heard at this office, before Thomas Leach, Esq. the sitting magistrate, against a man of the name of Edmund Rhodes, charged with having, on the 12th of August last, dyed, fabricated, and manufactured, divers large quantities, viz. one hundred weight of sloe leaves, one hundred weight of ash leaves, one hundred weight of elder leaves, and one hundred weight of the leaves of a certain other tree, in imitation of tea, contrary to the statute of the 17th of Geo. III. whereby the said Edmund Rhodes had, for every pound of such leaves so manufactured, forfeited the sum of 5*l.* making the total of the penalties amount to 2,000*l.* The second count in the information charged the said Rhodes with having in his possession the above quantity of sloe, ash, elder, and other leaves, under the like penalty of 2,000*l.* The third count charged him with having, on the said 12th of August last, in his possession, divers quantities, exceeding six pounds weight of each respective kind of leaves; viz. fifty pounds weight of green sloe leaves, fifty pounds weight of green leaves of ash, fifty pounds weight of green leaves of elder, and fifty pounds weight of the green leaves of a certain other tree; not having proved that such leaves were gathered with theconsent of the owners of the trees and shrubs from which they were taken, and that such leaves were gathered for some other use, and not for the purpose of manufacturing the same in imitation of tea; whereby he had forfeited for each pound weight, the sum of 5*l.* amounting in the whole to 1,000*l.*; and, in default of payment, in each case, subjected himself to be committed to the house of correction for not more than twelve months, nor less than six months.

Mr. Denton, who appeared for the defendant, who was absent, said that he was a very poor man, with a family of five children, and was only the servant of the real manufacturer, and an ignorant man from the country, put into the premises to carry on the business, without knowing what the leaves were intended for. By direction of Mr. Mayo, who conducted the prosecution, several barrels and bags, filled with the imitation tea, were then brought into the office, and a sample from each handed round. To the eye they seemed a good imitation of tea.

The defendant was convicted in the penalty of 500*l.* on the second count.

The Attorney-General against Palmer.—This was an action by the Attorney-General against the defendant, Palmer, charging him with having in his possession a quantity of sloe-leaves and white-thorn leaves, fabricated into an imitation of tea.

Mr. Dauncey stated the case to the jury, and observed that the defendant, Mr. Palmer, was a grocer. It would appear that a regular manufactory was established in Goldstone-street. The parties by whom the manufactory was conducted, was a person of the name of Proctor, and another person named J. Malins. They engaged others to furnish them with leaves, which, after undergoing a certain process, were sold to and drank by the public as tea. The leaves, in order to be converted into an article resembling black tea, were first boiled, then baked upon an iron plate; and, when dry, rubbed with the hand, in order to produce that curl which the genuine tea had. This was the most wholesome part of the operation; for the colour which was yet to be given to it, was produced by logwood. The green tea was manufactured in a manner more destructive to the constitution of those by whom it was drank. The leaves, being pressed and dried, were laid upon sheets of copper, where they received their colour from an article known by the name of Dutch pink. The article used in producing the appearance of the fine green bloom, observable on the China tea, was, however, decidedly a dead poison! He alluded to verdigris, which was added to the Dutch pink in order to complete the operation. This was the case which he had to bring before the jury; and hence it would appear, that, at the moment they were supposing they were drinking a pleasant and nutritious beverage, they were, in fact, in all probability, drinking the produce of the hedges round the metropolis, prepared for the purposes of deception in the most noxious manner. He trusted he should be enabled to trace to the possession of the defendant eighty pounds weight of the commodity he had been describing.

Thomas Jones deposed, that he knew Proctor, and was employed by him at the latter end of April, 1817, to gather black and white thorn leaves. Sloe leaves were the black thorn. Witness also knew John Malins, the son of William Malins, a coffee-roaster; he did not at first know the purpose for which the leaves were gathered, but afterwards learnt they were to make imitation tea. Witness did not gather more than one hundred and a half weight of these leaves; but he employed another person, of the name of John Bagster, to gather them. He had two-pence per pound for them. They were first boiled, and the water squeezed from them in a press. They were afterwards placed over a slow-fire upon sheets of copper to dry; while on the copper they were rubbed with the hand to curl them. At the time of boiling there was a little *verdigris* put into the water (this applied to green tea only.) After the leaves were dried, they were sifted, to separate the thorns and stalks. After they were sifted, more verdigris and some Dutch pink were added. The verdigris gave the leaves that green bloom observable on genuine tea.

The black tea went through a similar course as the green, except the application of Dutch pink: a little verdigris was put in the boiling, and to this was added a small quantity of logwood to dye it, and thus the manufacture was complete. The drying operation took place on sheets of iron. Witness knew the defendant, Edward Palmer; he took some of the mixture he had been describing, to his shop. The first time he took some was in May, 1817. In the course of that month, or the beginning of June, he took four or five seven-pound parcels; when he took it there, it was taken up to the top of the house. Witness afterwards carried some to Russell-street, which was taken to the top of the house, about one hundred weight and three quarters; from this quantity he carried fifty-three pounds weight to the house of the defendant's porter, by the desire of Mr. Malins; it was in paper parcels of seven pounds each.

John Bagster proved that he had been employed by Malins and Proctor, to gather sloe and white-thorn leaves: they were taken to Jones's house, and from thence to Malins' coffee-roasting premises; witness received two-pence per pound for them; he saw the manufacturing going on, but did not know much about it: witness saw the leaves on sheets of copper, in Goldstone-street.

This was the case for the Crown.—Verdict for the Crown, 840*l.*

The Attorney-General against John Prentice.—This was an information similar to the last, in which the defendant submitted to a verdict for the Crown.

The Attorney-General against Lawson Holmes.—In this case the defendant submitted to a verdict for the Crown.

The Attorney-General against John Orkney.—Thomas Jones proved that the defendant was a grocer, and in the month of May last he carried to his shop seven pounds of imitation tea, by the order of John Malins, for which he received the money, viz. 15*s.* 9*d.* or 2*s.* 3*d.* per pound.

The jury found a verdict for the Crown.—Penalties 70*l.*

The Attorney-General against James Gray.—The defendant submitted to a verdict for the Crown.—Penalties 120*l.*

The Attorney-General against H. Gilbert, and Powel.—These defendants submitted to a verdict.—Penalties 140*l.*

The Attorney-General against William Clarke.—This defendant also submitted to a verdict for the Crown.

The Attorney-General against George David Bellis.—This defendant submitted to a verdict for the Crown.

The Attorney-General against John Horner.—The defendant in this case was a grocer; it was proved by Jones that he received twenty pounds of imitation tea.—Verdict for the Crown.—Penalties 210*l.*

The Attorney-General against William Dowling.—This was a grocer. Jones proved that he delivered seven pounds of imitation tea at Mr.

Dowling's house, and received the money for it, namely 15*s*. 9*d*.—Penalties 70*l*.

METHOD OF DETECTING THE ADULTERATIONS OF TEA.

The adulteration of tea may be evinced by comparing the botanical characters of the leaves of the two respective trees, and by submitting them to the action of a few chemical tests.

The shape of the tea-leaf is slender and narrow, as shewn in this sketch, the edges are deeply serrated, and the end or extremity is acutely pointed. The texture of the leaf is very delicate, its surface smooth and glossy, and its colour is a lively pale green.

The sloe-leaf (and also the white-thorn leaf,) as shewn in this sketch, is more rounded, and the leaf is obtusely pointed. The serratures or jags on the edges are not so deep, the surface of the leaf is more uneven, the texture not so delicate, and the colour is a dark olive green.

These characters of course can be observed only after the dried leaves have been suffered to macerate in water for about twenty-four hours.

The leaves of some sorts of tea may differ in size, but the shape is the same in all of them; because all the different kinds of tea imported from China, are the produce of one species of plant, and the difference between the green and souchong, or black tea, depends chiefly upon the climate, soil, culture, age, and mode of drying the leaves.

Spurious black tea, slightly moistened, when rubbed on a sheet of white paper, immediately produces a blueish-black stain; and speedily affords, when thrown into cold water, a blueish-black tincture, which instantly becomes reddened by letting fall into it, a drop or two of sulphuric acid.

Two ounces of the suspected leaves, should be infused in half-a-pint of cold, soft water, and suffered to stand for about an hour. Genuine tea produces an amber-coloured infusion, which does not become reddened by sulphuric acid.

All the samples of spurious green tea (nineteen in number) which I have examined, were coloured with carbonate of copper (a poisonous substance,) and not by means of verdigris, or copperas. The latter substances would instantly turn the tea black; because both these metallic salts being soluble in water, are acted on by the astringent matter of the leaves, whether genuine or spurious, and convert the infusion into ink.

Tea, rendered poisonous by carbonate of copper, speedily imparts to liquid ammonia a fine sapphire blue tinge. It is only necessary to shake up in a stopped vial, for a few minutes, a tea-spoonful of the suspected leaves, with about two table-spoonsful of liquid ammonia, diluted with half its bulk of water. The supernatant liquid will exhibit a fine blue colour, if the minutest quantity of copper be present.

Green tea, coloured with carbonate of copper, when thrown into water impregnated with sulphuretted hydrogen gas, immediately acquires a black colour. Genuine green tea suffers no change from the action of these tests.

The presence of copper may be further rendered obvious, by mixing one part of the suspected tea-leaves, reduced to powder, with two or three parts of nitrate of potash, (or with two parts of chlorate of potash,) and projecting this mixture by small portions at a time, into a platina, or porcelain-ware crucible, kept red-hot in a coal fire; the whole vegetable matter of the tea leaves will thus become destroyed, and the oxide of copper left behind, in combination with the potash, of the nitrate of potash (or salt-petre,) or with the muriate of potash, if chlorate of potash has been employed.

If water, acidulated with nitric acid, be then poured into the crucible to dissolve the mass, the presence of the copper may be rendered manifest by adding to the solution, liquid ammonia, in such quantity that the pungent odour of it predominates.

Counterfeit Coffee.

The fraud of counterfeiting ground coffee by means of pigeon's beans and pease, is another subject which, not long ago, arrested the attention of the public: and from the numerous convictions of grocers prosecuted for the offence, it is evident that this practice has been carried on for a long time, and to a considerable extent.

The following statement exhibits some of the prosecutions, instituted by the Solicitor of the Excise, against persons convicted of the fraud of manufacturing spurious, and adulterating genuine coffee.

Alexander Brady, a grocer, prosecuted and convicted of selling *sham-coffee*, said, "I have sold it for twenty years." Some of the persons prosecuted by the Solicitor of the Excise for this fraud, we might, at first sight, be inclined to believe, were inconscious that the adulterating of genuine coffee with spurious substances was illegal; but this ignorance affords no excuse, explicitly states: "If after the first day of September, 1803, any burnt, scorched, or roasted pease, beans, or other grain, or vegetable substance or substances prepared or manufactured for the purpose of being in imitation of or in any respect to resemble coffee or cocoa, or to serve as a substitute for coffee or cocoa, or alleged or pretended by the possessor or vender thereof so to be, *shall be made,* or kept for sale, or shall be *offered* or *exposed to sale,* or shall be *found* in the custody or possession of any*dealer* or dealers in or *seller* or sellers of *coffee*, or if any burnt, scorched, or roasted pease, beans, or other grain, or vegetable substance or substances not being coffee, shall be called by the preparer, manufacturer, possessor, or vender thereof, by the name of *English* or *British* coffee, or *any other name* of coffee, or by the name of *American* cocoa, or *English* or *British*cocoa, or any other name of cocoa, the same respectively shall be forfeited, together with the packages containing the same, and shall and may be seized by any officer or officers of Excise; and the person or persons preparing, manufacturing, or selling the same, or having the same in his, her, or their custody or possession, or the dealer or dealers in or seller or sellers of coffee or cocoa, in whose custody the same shall be found, shall forfeit and lose the sum of one hundred pounds."

The Attorney-General against William Malins.—This was an information filed by the Attorney-General against the defendant, charging him, he being a dealer in coffee, with having in his possession a large quantity of imitation coffee, made from scorched pease and beans, resembling coffee, and intended to be sold as such, contrary to the statute of the 43d of the King, whereby he became liable to pay a fine of 100*l.*

J. Lawes deposed that he had lived servant with the defendant; he constantly roasted pease and beans, and ground them into powder. When so ground, the powder very much resembled coffee. Sometimes the sweepings of the coffee were thrown in among the pease and beans. Witness carried out this powder to several grocers in different parts of the town.

Thomas Jones lived with the defendant. His occupation was roasting and grinding pease and beans. They looked, when ground, the same as coffee. Witness had seen Mr. John Malins sweep up the refuse coffee, and mix it with the pease and beans. He had taken out this mixture to grocers.

J. Richardson, an excise-officer, deposed, that, in December 1817, he went to the premises of the defendant, and there seized four sacks, five tubs, and nine pounds in paper, of a powder made to resemble coffee. The quantity ground was 1,567 pounds; it had all the appearance of coffee; and a little coffee being mixed with it, any common person might be deceived. He also seized two sacks, containing 279 pounds of whole pease and beans roasted. Among the latter were some grains of coffee. The witness here produced samples of the articles seized.

John Lawes deposed, that the articles exhibited were such as he was in the habit of manufacturing while in Mr. Malins' employment.

The jury found a verdict for the Crown.—Penalty 100*l.*

The King against Chaloner.—Mr. Chaloner, a dealer in tea and coffee, was charged on the oaths of Charles Henry Lord and John Pearson, both Excise officers, with having in his possession, on the 17th of March, nine pounds of spurious coffee, consisting of burnt pease, beans, and gravel or sand, and a portion of coffee, and with selling some of the same; also with having in his possession seventeen pounds of vegetable powder, and an article imitating coffee, which contained not a particle of genuine coffee.

The defendant was convicted in the penalty of 90*l*.

The King against Peether.—This was an information against Mr. Thomas Peether, tea and coffee dealer, charging him with having in his possession a quantity of imitation coffee (or vegetable powder) on the 25th of April last.

The case being proved by the evidence of several witnesses, the defendant was convicted in the penalty of 50*l*.

The King against Topping.—This was an information against Mr. John Lewis Topping, a dealer in tea and coffee, charging him with having thirty-seven pounds of vegetable powder in his possession. The article seized was produced to the commissioners of the Excise.

The defendant was convicted in the penalty of 50*l*.

The King against Samuel Hallett.—The defendant, Hallett, a grocer and dealer in tea and coffee, was charged with having seven pounds of imitation coffee in his possession.

Charles Henry Lord, an officer of the Excise, being sworn, stated, that he and Spencer, an officer, went, on the 28th of February last, to the shop of the defendant, and asked for an ounce of coffee, at three halfpence per ounce. He received the same, and having paid for it, left the shop. He examined the article, and found it was part coffee, and part imitation coffee, or what the defendant called vegetable powder, which is nothing more nor less than burnt pease and beans ground in a mill.

Spencer, the officer of the Excise, corroborated the above evidence, and stated, that the sham-coffee seized at the defendant's house was shown to Mr. Joseph Hubbard, grocer, and tea and coffee dealer, in High-street, in the Borough of Southwark.

Mr. Hubbard being sworn, stated, that he had examined the sham-coffee seized by the officers in the defendant's shop. The one ounce purchased by Lord, he knew to be nothing else than black pigeon's beans; there was no coffee amongst it.

The defendant was convicted in the penalty of 50*l.*

The King against Fox.—Mr. Edward Fox, grocer, and dealer in tea and coffee, was charged with having a large quantity of sham-coffee in his possession, and with selling the same for genuine coffee.

Henry Spencer, an officer of the Excise, stated, that on the 21st of February he and Lord, another officer, went to the defendant's shop and purchased an ounce of coffee, for which he paid three halfpence. They examined it, and he was satisfied it was not genuine coffee; they purchased another ounce (which he produced to the commissioners of the Excise, who examined it); they were convinced it consisted partly of coffee and beans and pease.

The defendant, in his defence said, that the poor people wanted a low-price article; and by mixing the vegetable powder and coffee together, he was able to sell it at three halfpence an ounce; he had sold it for years; he did it as a matter of accommodation to the poor, who could not give a higher price; he did not sell it for genuine coffee.

Commissioner.—"Then you have been defrauding the public for many years, and injuring the revenue by your illicit practices: the poor have an equal right to be supplied with as genuine an article as the rich."

He was convicted in the penalty of 50*l.*

The King against Brady.—The defendant, Mr. Alexander Brady, grocer, and dealer in tea and coffee, was charged with having, on the 28th of February last, in his possession eighteen pounds of sham-coffee, and selling the same for genuine coffee.

Lord and Pearson, Excise officers, stated, that they purchased an ounce of coffee of the defendant, on the 28th of February, and upon examining it they discovered that it was made up of pease and beans, ground with a small quantity of coffee. They also found eighteen pounds of vegetable powder mixed with coffee, in a state prepared for sale, wrapped in papers.

One of the commissioners tasted some of the eighteen pounds of sham-coffee produced by the officers, and declared that it was a most infamous stuff, and unfit for human food.

Defendant.—"Why, I have sold it for twenty years."

Commissioner.—"Then you have been for twenty years acting most dishonestly, defrauding the revenue; and the health of the poor must have suffered very much by taking such an unwholesome article. Your having dealt in this article so long aggravates your case; you have for twenty years been selling burnt beans and pease for genuine coffee.—You are convicted in the penalty of 50*l.*"

The King against Bowser.—The excise officers stated, that on the 28th of February they went to his shop: he was a grocer, dealer in tea and coffee; they seized seven pounds and a half of vegetable powder, which contained very little coffee, if any; and also a quarter of a pound of coffee mixed with vegetable powder.

The defendant pleaded guilty to the charge, and prayed the court to mitigate the penalty. He was convicted in the penalty of 50*l.*

The King against Thomas Owen.—The defendant, an extensive dealer in tea and coffee, appeared to an information charging him with having in his possession, and selling, a quantity of deleterious ingredients, and mixing them with coffee.

Charles Henry Lord deposed, that on the 26th of February, he found, at the shop of the defendant, nineteen pounds of a composition consisting of beans and pease ground, and prepared so as to imitate coffee. He also discovered two pounds and a half of a mixture of coffee and vegetable powder. On the same day he proceeded to another shop of the defendant, and he there found five pounds more of the same stuff.

Samples of the composition, in its mixed and unmixed state, were produced.

Mr. Lawes addressed the commissioners on behalf of the defendant, in mitigation of punishment; for he did not mean to deny the offence. His client was a very young man, and had been most unfortunate in business. He was not aware until lately of the existence of any law by which it could be punished.

The Commissioners observed, that they had a double duty to perform, namely, to protect the revenue from fraud, and to prevent the public from being imposed upon and injured by ingredients served to them instead of the food they intended to purchase. The fraud upon the revenue was, in the estimation of the court, the least part of the offence. Under all the circumstances, however, the court was inclined to be lenient to the defendant.

He was convicted in the penalty of 50*l.* for each quantity of sham-coffee.

Mr. Greely and Mr. William Dando were fined 20*l.* each; and Mr. Hirling and Mr. Terry were fined 90*l.* each for selling spurious coffee.

The adulteration of ground coffee, with pease and beans, is beyond the reach of chemical analysis; but it may, perhaps, not be amiss on this occasion to give to our readers a piece of advice given by a retired grocer to a friend, at no distant period:—"Never, my good fellow," he said, "purchase from a grocer any thing which passes through his mill. You know not what you get instead of the article you expect to receive—coffee, pepper, and all-spice, are all mixed with substances which detract from their own natural qualities."—Persons keeping mills of their own can at all times prevent these impositions.

By the Excise laws at present existing in this country, the various degrees of strength of brandy, rum, arrack, gin, whiskey, and other spiritous liquors, chiefly composed of little else than spirit of wine, are determined by the quantity of alcohol of a given specific gravity contained in the spiritous liquors of a supposed unknown strength. The great public importance of this subject in this country, where the consumption of spiritous liquors adds a vast sum to the public revenue, has been the means of instituting many very interesting series of experiments on this subject. The instrument used for that purpose by the Customs and officers of Excise, is called *Sikes*'s hydrometer, which has now superseded the instrument called *Clark*'s hydrometer, heretofore in use.

The specific gravity or strength of the legal standard spirit of the Excise, is technically called *proof* or *proof spirit*. "This liquor (not being spirit sweetened, or having any ingredient dissolved in it, to defeat the strength thereof,) at the temperature of 57° Faht. weighs exactly 12/13th parts of an equal measure of distilled water;" and with this spirit the strength of all other spiritous liquors are compared according to law.

The strength of spirit stronger than *proof* or *over proof*, as it is termed by the revenue officers, is indicated by the bulk of water necessary to reduce a given volume of it, to the legal standard spirit, denominated *proof*—namely; if one gallon of water be required to bring twenty gallons of brandy, rum, or any other spirit, to proof, that spirit is said to be *1 to 20 over proof*. If one gallon of water be required to bring 15, 10, 5, or 2 gallons of the liquor to *proof*, it is said to be 1 to 15, 1 to 10, 1 to 5, and 1 to 2, *over proof*.

The strength of brandy, rum, arrack, gin, or other spiritous liquors, weaker than *proof*, or under *proof*, is estimated by the quantity of water which would be necessary to abstract or bring the spirit up to proof.

Thus, if from twenty gallons of brandy one gallon of water must be abstracted to bring it to proof, it is said to be 1 in 20 under proof. If from 15, 10, 5, or 2 gallons of the liquor, 1 gallon of water must be abstracted to bring it to proof, it is said to be 1 in 15, 1 in 10, 1 in 5, and 1 in 2 under proof.

It is necessary to understand this absurd language, which is in use amongst the officers of Excise and dealers in spirit, in order to know what is meant in commerce by the strength of spiritous liquors of different denominations. And hence, for the business of the exciseman, a table has been constructed, expressing the strength or specific gravity of mixtures of different proportions of spirit and water, at different degrees of temperature; and according to this table the duty on spirit is now levied.

Brandy and rum is seizable, if sold by, or found in the possession of, the dealer, unless it possesses a certain strength. The following are the words of the Act:

"No distiller, rectifier, compounder or dealer, shall serve or send out any foreign spirits, of a lower strength than that of 1 in 6 under hydrometer proof, nor have in his possession any foreign spirits mixed together, except shrub, cherry or raspberry brandy, of lower strength than as aforesaid, upon pain of such spirits being forfeited; and such spirits, with the casks and vessels containing the same, may be seized by any officer of Excise."

We have, therefore, a ready check against the frauds of the dishonest dealers, in spiritous liquors. If the spirit merchant engages to deliver a liquor of a certain strength, the hydrometer is by far the most easy and expeditious check that can be adopted to guard against frauds of receiving a weaker liquor for a stronger one; and to those individuals who are in the habit of purchasing large quantities of brandy, rum, or other spiritous liquors, the hydrometer renders the greatest service. For it is by no means an uncommon occurrence to meet with brandy, rum, and other spiritous liquors, of a specific gravity very much below the pretended strength which the liquor ought to possess.

The following advice, given to his readers, by the author of a Treatise on Brewing and Distilling, may serve to put the unwary on their guard against some of the frauds practised by mercenary dealers.

"It is a custom among retailing distillers, which I have not taken notice of in this directory, to put one-third or one-fourth part of proof molasses brandy, proportionably, to what rum they dispose of; which cannot be distinguished, but by an extraordinary palate, and does not at all lessen the body or proof of the goods; but makes them about two shillings a gallon cheaper; and must be well mixed and incorporated together in your retailing cask; but you should keep some of the best rum, not adulterated, to please some customers, whose judgment and palate must be humoured."

"When you are to draw a sample of goods to shew a person that has judgment in the proof, do not draw your goods into a phial to be tasted, or make experiment of the strength thereof that way, because the proof will not hold except the goods be exceedingly strong; but draw the pattern of goods rather into a glass from the cock, to run very small, or rather draw off a small quantity into a little pewter pot and pour it into your glass, extending your pot as high above the glasses as you can without wasting it, which makes the goods carry a better head abundantly, than if the same goods were to be put and tried in a phial."

"You must be so prudent as to make a distinction of the persons you have to deal with; what goods you sell to gentlemen for their own use, who require a great deal of attendance, and as much for time of payment, you must take a considerably greater price than of others; what goods you sell to persons where you believe there is a manifest, or at least some hazard of your money, you may safely sell for more than common profit; what goods you sell to the poor, especially medicinally, (as many of your goods are sanative,) be as compassionate as the cases require."

"All brandies, whether French, Spanish, or English; being proof goods, will admit of one point of *liquor* to each gallon, to be made up and incorporated therewith in your cask, for retail, or selling smaller quantities; and all persons that insist upon having proof goods, which not one in twenty understands, you must supply out of what goods are not so reduced, though at a higher price."

Such is the advice given by Mr. Shannon.

The mode of judging by the taste of spiritous liquors is deceitful. A false strength is given to a weak liquor, by infusing in it acrid vegetable substances, or by adding to it a tincture of grains of paradise and Guinea pepper. These substances impart to weak brandy or rum, an extremely hot and pungent taste.

Brandy and rum is also frequently sophisticated with British molasses, or sugar-spirit, coloured with burnt sugar.

The flavour which characterises French brandy, and which is owing to a small portion of a peculiar essential oil contained in it, is imitated by distilling British molasses-spirit over wine lees; but the spirit, prior to being distilled over wine lees, is previously deprived, in part, of its peculiar disagreeable flavour, by rectification over fresh burnt charcoal and quick-lime. Other brandy-merchants employ a spirit obtained from raisin wine, which is suffered to pass into an incipient ascescency. The spirit thus procured partakes strongly of the flavour which is characteristic to foreign brandy.

Oak saw-dust, and a spiritous tincture of raisin stones, are likewise used to impart to new brandy and rum a *ripe taste*, resembling brandy or rum long kept in oaken casks, and a somewhat oily consistence, so as to form a durable froth at its surface, when strongly agitated in a vial. The colouring substances are burnt sugar, or molasses; the latter gives to imitative brandy a luscious taste, and fulness *in the mouth*. These properties are said to render it particularly fit for the retail London customers.

The following is the method of compounding or *making up*, as it is technically called, *brandy* for retail:

	Gallons
"To ten puncheons of brandy	1081
Add flavoured raisin spirit	118
Tincture of grains of paradise	4
Cherry laurel water	2
Spirit of almond cakes	2

— — —

1207

"Add also 10 handfuls of oak saw-dust; and give it *complexion* with burnt sugar."

METHOD OF DETECTING THE ADULTERATIONS OF BRANDY, RUM, AND MALT SPIRIT.

The false strength of brandy or rum is rendered obvious by diluting the suspected liquor with water; the acrimony of the capsicum, and grains of paradise, or pepper, may then be readily discovered by the taste.

The adulteration of brandy with British molasses, or sugar-spirit, becomes evident by rubbing a portion of the suspected brandy between the palms of the hands; the spirit, as it evaporates, leaves the disagreeable flavour which is peculiar to all British spirits. Or the liquor may be deprived of its alcohol, by heating a portion in a spoon over a candle, till the vapour ceases to catch fire on the approach of a lighted taper. The residue thus obtained, of genuine French brandy, possesses a vinous odour, still resembling the original flavour of the brandy, whilst the residue, produced from sophisticated brandy, has a peculiarly disagreeable smell, resembling gin, or the breath of habitual drunkards.

Arrack is coarsely imitated by adding to rum a small quantity of pyroligneous acid and some flowers (acid) of benzoe. The compound thus produced, however, must be pronounced a bad one. The author of a very popular Cookery Book, directs two scruples of benzoic acid to be dissolved in one quart of rum, to make "*mock arrack.*"

MALT SPIRIT.

Malt spirit, or gin, the favourite liquor of the lower order of people, which is characterised by the peculiar flavour of juniper berries, over which the raw spirit is distilled, is usually obtained from a mixture of malt and barley: sometimes both molasses and corn are employed, particularly if there be a scarcity of grain. But the flavour of whiskey, which is made from barley and oats, is owing to the malted grain being dried with peat, the smoke of which gives it the characteristic taste.

The malt distiller is not allowed to furnish, under a heavy penalty, any crude or raw spirit to the rectifier or manufacturer of gin, of a greater strength than seven per cent. over proof. The rectifier who receives the spirit from the malt distiller is not allowed, under a certain penalty, to sweeten the liquor with sugar or other substances; nor is he permitted to send out the spirit to his customers but of a certain strength, as is obvious from the following words of the Act:

"No rectifier or compounder shall sell or send out any British brandy, British rectified spirits, British compounds, or other British spirits, of greater strength than that of one in five under hydrometer proof: and if he shall sell and send out any such spirits of a greater strength than that of one in five under hydrometer proof, such spirits, with the casks or vessels containing the same, shall be forfeited, and may be seized by any officer of Excise; and he shall also forfeit treble the value of such spirit, or 50*l.* at the election of the King's attorney-general, or the person who shall sue for the same; the single value of such spirits to be estimated at the highest London Price."

If we examine gin, as retailed, we shall soon be convinced that it is a custom, pretty prevalent amongst dealers, to weaken this liquor considerably with water, and to sweeten it with sugar. This fraud may readily be detected by evaporating a quantity of the liquor in a table-spoon over a candle, to dryness; the sugar will thus be rendered obvious, in the form of a gum-like substance, when the spirit is volatilised.

One hundred and twenty gallons of genuine gin, as obtained from the wholesale manufactories, are usually *made up* by fraudulent retailers, into a saleable commodity, with fourteen gallons of water and twenty-six pounds of sugar. Now this dilution of the liquor produces a turbidness; because the oil of juniper and other flavouring substances which the spirit holds in solution, become precipitated by virtue of the water, and thus cause the liquor to assume an opaline colour: and the spirit thus weakened, cannot readily be rendered clear again by subsidence. Several expedients are had recourse to, to clarify the liquor in an expeditious manner; some of which are harmless; others are criminal, because they render the liquor poisonous.

One of the methods, which is innocent, consists in adding to the weakened liquor, first, a portion of alum dissolved in water, and then a solution of sub-carbonate of potash. The whole is stirred together, and left undisturbed for twenty-four hours. The precipitated alumine thus produced from the alum, by virtue of the sub-carbonate of potash, acts as a strainer upon the milky liquor, and carries down with it the finely divided oily matter which produced the blue colour of the diluted liquor. Roach, or Roman alum, is also employed, without any other addition, for clarifying spiritous liquors.

"*To reduce unsweetened Gin.*

"A tun of fine gin	252	gallons
"Water	36	
— —		
"Which added together make	288	gallons
"The *doctor is now put* on, and it is further reduced with water	19	
— —		
"Which gives Total	307	gallons of gin.

"This done, let 1 lb. of alum be just covered with water, and dissolved by boiling; rummage the whole well together, and pour in the alum, and the whole will be fine in a few hours."

"*To prepare and sweeten British Gin.*

"Get from your distiller an empty puncheon or cask, which will contain about 133 gallons. Then take a cask of clear rectified spirits,

120 gallons, of the usual strength as rectifiers sell their goods at, put the 120 gallons of spirits into your empty cask.

"Then take a quarter of an ounce of oil of vitriol, half an ounce of oil of almonds, a quarter of an ounce of oil of turpentine, one ounce of oil of juniper berries, half a pint of spirit of wine, and half a pound of lump sugar. Beat or rub the above in a mortar. When well rubbed together, have ready prepared half a gallon of lime water, one gallon of rose water; mix the whole in either a pail, or cask, with a stick, till every particle shall be dissolved; then add to the foregoing, twenty-five pounds of sugar dissolved in about nine gallons of rain or Thames water, or water that has been boiled, mix the whole well together, and stir them carefully with a stick in the 133 gallons cask.

"To *force down* the same, take and boil eight ounces of alum in three quarts of water, for three quarters of an hour; take it from the fire, and dissolve by degrees six or seven ounces of salt of tartar. When the same is milk-warm pour it into your gin, and stir it well together, as before, for five minutes, the same as you would a butt of beer newly fined. Let your cask stand as you mean to draw it. At every time you purpose to sweeten again, that cask must be well washed out; and take great care never to shake your cask all the while it is drawing."

Another method of fining spiritous liquors, consists in adding to it, first, a solution of sub-acetate of lead, and then a solution of alum. This practice is highly dangerous, because part of the sulphate of lead produced, remains dissolved in the liquor, which it thus renders poisonous. Unfortunately, this method of clarifying spiritous liquors, I have good reason to believe, is more frequently practised than the preceding method, because its action is more rapid; and it imparts to the liquor a fine *complexion*, or great refractive power; hence some vestiges of lead may often be detected in malt spirit.

The weakened spirit is then sweetened with sugar, and, to cover the raw taste of the malt spirit, *false strength* is given to it with grains of paradise, Guinea pepper, capsicum, and other acrid and aromatic substances.

METHOD OF DETECTING THE PRESENCE OF LEAD IN SPIRITOUS LIQUORS.

The presence of lead may be detected in spiritous liquors, as stated. The cordial called shrub frequently exhibits vestiges of copper. This contamination, I have been informed, is accidental, and originates from the metallic vessels employed in the manufacture of the liquor.

METHOD OF ASCERTAINING THE QUANTITY OF ALCOHOL IN DIFFERENT KINDS OF SPIRITOUS LIQUORS.

The quantity of real alcohol in any spiritous liquors may readily be ascertained by simple distillation, which process separates the alcohol from the water and foreign matters contained in the liquor. Put any quantity of brandy, rum, or malt spirit diluted with about one-fourth its bulk of water, into a retort fitted to a capacious receiver, and distil with a gentle heat. The strongest spirit distils over first into the receiver, and the strength of the obtained products decreases, till at last it contains so much water as no longer to be inflammable by the approach of a lighted taper, when held in a spoon over a candle If the process be continued, the distilled product becomes milky, scarcely spiritous to the smell, and of an acidulous taste. The distilling operation may then be discontinued. If the first, fourth or third part of the distilled product has been set apart, it will be found a moderately strong alcohol, and the remainder one more diluted. If the whole distilled spirit be mixed with perfectly dry subcarbonate of potash, the alcohol will float at the top of the potash, as stated, it will separate into two distinct fluids. If the decanted alcohol be redistilled carefully with a very gentle heat, over a small portion of dry quick lime, or muriate of lime, it will be obtained extremely pure, and of a specific gravity of about 825, at 60° of temperature. Its flavour will vary according to the kind of spiritous liquor from which it is obtained.

Table exhibiting the Per Centage of Alcohol (of 825 specific gravity) contained in various kinds of spiritous Liquors.

	Proportion of Alcohol per Cent. by Measure.
Brandy, Cogniac, average proportion of 4 samples	52,75
Ditto, Bourdeaux, ditto ditto	54,50
Ditto, Cette	53,00
Ditto, Naples, average of 3 samples	53,25
Ditto, Spanish average of 6 samples	52,28
Rum	53,68
Ditto, Leeward, average of 9 samples	53,00
Scotch Whiskey, average of 6 samples	53,50
Irish Ditto, average of 4 samples	54,25
Arrack, Batavia	49,50
Dutch Geneva	52,25
Gin (Hodges's,) 3 samples, procured from retail dealers	48,25
Ditto (Ditto,) procured from the manufacturer	52,35

Poisonous Cheese.

Several instances have come under my notice in which Gloucester cheese has been contaminated with red lead, and has produced serious consequences on being taken into the stomach. In one poisonous sample which it fell to my lot to investigate, the evil had been caused by the sophistication of the anotta, employed for colouring cheese. This substance was found to contain a portion of red lead; a method of sophistication which has lately been confirmed by the following fact, communicated to the public by Mr. J. W. Wright, of Cambridge.

"As a striking example of the extent to which adulterated articles of food may be unconsciously diffused, and of the consequent difficulty of detecting the real fabricators of them, it may not be uninteresting to relate to your readers, the various steps by which the fraud of a poisonous adulteration of cheese was traced to its source.

"Your readers ought here to be told, that several instances are on record, that Gloucester and other cheeses have been found contaminated with red lead, and that this contamination has produced serious consequences. In the instance now alluded to, and probably in all other cases, the deleterious mixture had been caused ignorantly, by the adulteration of the anotta employed for colouring the cheese. This substance, in the instance I shall relate, was found to contain a portion of red lead; a species of adulteration which subsequent experiments have shewn to be by no means uncommon. Before I proceed further to trace this fraud to its source, I shall briefly relate the circumstance which gave rise to its detection.

"A gentleman, who had occasion to reside for some time in a city in the West of England, was one night seized with a distressing but indescribable pain in the region of the abdomen and of the stomach, accompanied with a feeling of tension, which occasioned much restlessness, anxiety, and repugnance to food. He began to apprehend the access of an inflammatory disorder; but in twenty-four hours the symptoms entirely subsided. In four days afterwards he experienced an attack precisely similar; and he then recollected, that having, on both occasions, arrived from the country late in the evening, he had ordered a plate of toasted Gloucester cheese, of which he had partaken heartily; a dish which, when at home, regularly served him for supper. He attributed his illness to the cheese. The circumstance was mentioned to the mistress of the inn, who expressed great surprise, as the cheese in question was not purchased from a country dealer, but from a highly respectable shop in London. He, therefore, ascribed the before-mentioned effects to some peculiarity in his constitution. A few days afterwards he partook of the same cheese; and he had scarcely retired to rest, when a most violent cholic seized him, which lasted the whole night and part of the ensuing day. The cook was now directed henceforth not to serve up any toasted cheese, and he never again experienced these distressing symptoms. Whilst this matter was a subject of conversation in the house, a servant-maid mentioned that a kitten had been violently sick after having eaten the rind cut off from the cheese prepared for the gentleman's supper. The landlady, in consequence of this statement, ordered the cheese to be examined by a chemist in the vicinity, who returned for answer, that the cheese was contaminated with lead! So unexpected an answer arrested general attention, and more particularly as the suspected cheese had been served up for several other customers.

"Application was therefore made by the London dealer to the farmer who manufactured the cheese: he declared that he had bought the anotta of a mercantile traveller, who had supplied him and his neighbours for years with that commodity, without giving occasion to a single complaint. On subsequent inquiries, through a circuitous channel, unnecessary to be detailed here at length, on the part of the manufacturer of the cheese, it was found, that as the supplies of anotta had been defective and of inferior quality, recourse had been had to the expedient of colouring the commodity with vermilion. Even this admixture could not be considered deleterious. But on further application being made to the druggist who sold the article, the answer was, that the vermilion had been mixed with a portion of red lead; and the deception was held to be perfectly innocent, as frequently practised on the supposition, that the vermilion would be used only as a pigment for house-painting. Thus the druggist sold his vermilion in the regular way of trade, adulterated with red lead to increase his profit, without any suspicion of the use to which it would be applied; and the purchaser who adulterated the anotta, presuming that the vermilion was genuine, had no hesitation in heightening the colour of his spurious anotta with so harmless an adjunct. Thus, through the circuitous and diversified operations of commerce, a portion of deadly poison may find admission into the necessaries of life, in a way which can attach no criminality to the parties through whose hands it has successively passed."

This dangerous sophistication may be detected by macerating a portion of the suspected cheese in water impregnated with sulphuretted hydrogen, acidulated with muriatic acid; which will instantly cause the cheese to assume a brown or black colour, if the minutest portion of lead be present.

Counterfeit Pepper.

Black pepper is the fruit of a shrubby creeping plant, which grows wild in the East Indies, and is cultivated, with much advantage, for the sake of its berries, in Java and Malabar. The berries are gathered before they are ripe, and are dried in the sun. They become black and corrugated on the surface.

This factitious pepper-corns have of late been detected mixed with genuine pepper, is a fact sufficiently known. Such an adulteration may prove, in many instances of household economy, exceedingly vexatious and prejudicial to those who ignorantly make use of the spurious article. I have examined large packages of both black and white pepper, by order of the Excise, and have found them to contain about 16 per cent. of this artificial compound. The spurious pepper is made up of oil cakes (the residue of lintseed, from which the oil has been pressed,) common clay, and a portion of Cayenne pepper, formed in a mass, and granulated by being first pressed through a sieve, and then rolled in a cask. The mode of detecting the fraud is easy. It is only necessary to throw a sample of the suspected pepper into a bowl of water; the artificial pepper-corns fall to powder, whilst the true pepper remains whole.

Ground pepper is very often sophisticated by adding to a portion of genuine pepper, a quantity of pepper dust, or the sweepings from the pepper warehouses, mixed with a little Cayenne pepper. The sweepings are known, and purchased in the market, under the name of P. D. signifying pepper dust. An inferior sort of this vile refuse, or the sweepings of P. D. is distinguished among venders by the abbreviation of D. P. D. denoting, dust (dirt) of pepper dust.

The adulteration of pepper, and the making and selling commodities in imitation of pepper, are prohibited, under a severe penalty. The following are the words of the Act:

"And whereas commodities made in imitation of pepper have of late been sold and found in the possession of various dealers in pepper, and other persons in Great Britain; be it therefore enacted, that from and after the said 5th day of July, 1819, if any commodity or substance shall be prepared by any person in imitation of pepper, shall be mixed with pepper, or sold or delivered as and for, or as a substitute for, pepper, or if any such commodity or substance, alone or mixed, shall be kept for sale, sold, or delivered, or shall be offered or exposed to sale, or shall be in the custody or possession of any dealer or seller of pepper, the same, together with all pepper with which the same shall be mixed, shall be forfeited, with the packages containing the same, and shall and may be seized by any officer of excise; and the person preparing, manufacturing, mixing as aforesaid, selling, exposing to sale, or delivering the same, or having the same in his, her, or their custody or possession, shall forfeit the sum of one hundred pounds."

WHITE PEPPER.

The common white pepper is factitious, being prepared from the black pepper in the following manner:—The pepper is first steeped in sea water and urine, and then exposed to the heat of the sun for several days, till the rind or outer bark loosens; it is then taken out of the steep, and, when dry, it is rubbed with the hand till the rind falls off. The white fruit is then dried, and the remains of the rind blown away like chaff. A great deal of the peculiar flavour and pungent hot taste of the pepper is taken off by this process. White pepper is always inferior in flavour and quality to the black pepper.

However, there is a sort of native white pepper, produced on a species of the pepper plant, which is much better than the factitious, and indeed little inferior to the common black pepper.

Poisonous Cayenne Pepper.

Cayenne pepper is an indiscriminate mixture of the powder of the dried pods of many species of capsicum, but especially of the capsicum frutescens, or bird pepper, which is the hottest of all.

This annual plant, a native of South America, is cultivated in large quantities in our West-India islands, and even frequently in our gardens, for the beauty of its pods, which are long, pointed, and pendulous, at first of a green colour, and, when ripe, of a bright orange red. They are filled with a dry loose pulp, and contain many small, flat, kidney-shaped seeds. The taste of capsicum is extremely pungent and acrimonious, setting the mouth, as it were, on fire.

The principle on which its pungency depends, is soluble in water and in alcohol.

It is sometimes adulterated with red lead, to prevent it becoming bleached on exposure to light. This fraud may be readily detected by shaking up part of it in a stopped vial containing water impregnated with sulphuretted hydrogen gas, which will cause it speedily to assume a dark muddy black colour. Or the vegetable matter of the pepper may be destroyed, by throwing a mixture of one part of the suspected pepper and three of nitrate of potash (or two of chlorate of potash) into a red-hot crucible, in small quantities at a time. The mass left behind may then be digested in weak nitric acid, and the solution assayed for lead by water impregnated with sulphuretted hydrogen.

Poisonous Pickles.

Vegetable substances, preserved in the state called pickles, by means of the antiseptic power of vinegar, whose sale frequently depends greatly upon a fine lively green colour; and the consumption of which, by sea-faring people in particular, is prodigious, are sometimes intentionally coloured by means of copper. Gerkins, French beans, samphires, the green pods of capsicum, and many other pickled vegetable substances, oftener than is perhaps expected, are met with impregnated with this metal. Numerous fatal consequences are known to have ensued from the use of these stimulants of the palate, to which the fresh and pleasing hue has been imparted according to the deadly *formulæ* laid down in some modern cookery books, such as boiling the pickles with half-pence, or suffering them to stand for a considerable period in brazen vessels.

Dr. Percival has given an account of "a young lady who amused herself, while her hair was dressing, with eating samphire pickles impregnated with copper. She soon complained of pain in the stomach; and, in five days, vomiting commenced, which was incessant for two days. After this, her stomach became prodigiously distended; and, in nine days after eating the pickles, death relieved her from her suffering."

Among many recipes which modern authors of cookery books have given for imparting a green colour to pickles, the following are particularly deserving of censure; and it is to be hoped that they will be suppressed in future editions of the works.

"*To Pickle Gerkins.*—"Boil the vinegar in a bell-metal or copper pot; pour it boiling hot on your cucumbers."

"*To make greening.*—"Take a bit of verdigris, the bigness of a hazel-nut, finely powdered; half-a-pint of distilled vinegar, and a bit of alum powder, with a little bay salt. Put all in a bottle, shake it, and let it stand till clear. Put a small tea-spoonful into codlings, or whatever you wish to green."

Mrs. E. Raffald directs, "to render pickles green, boil them with halfpence, or allow them to stand for twenty-four hours in copper or brass pans."

To detect the presence of copper, it is only necessary to mince the pickles, and to pour liquid ammonia, diluted with an equal bulk of water, over them in a stopped phial: if the pickles contain the minutest quantity of copper, the ammonia assumes a blue colour.

Adulteration of Vinegar.

Vinegar, as prepared in this country, from malt, should be of a pale brown colour, perfectly transparent, of a pleasant, somewhat pungent, acid taste, and fragrant odour, but without any acrimony. From the mucilaginous impurities which malt vinegar always contains, it is apt, on exposure to air, to become turbid and ropy, and at last vapid. The inconvenience is best obviated by keeping the vinegar in bottles completely filled and well corked; and it is of advantage to boil it in the bottles a few minutes before they are corked.

Vinegar is sometimes largely adulterated with sulphuric acid, to give it more acidity. The presence of this acid is detected, if, on the addition of a solution of acetate of barytes, a white precipitate is formed, which is insoluble in nitric acid, after having been made red-hot in the fire. With the same intention, of making the vinegar appear stronger, different acrid vegetable substances are infused in it. This fraud is difficult of detection; but when tasted with attention, the pungency of such vinegar will be found to depend rather on acrimony than acidity.

Distilled vinegar, which is employed for various purposes of domestic economy, is frequently distilled, not in glass, as it ought to be, but in common stills with a pewter pipe, whence it cannot fail to acquire a metallic impregnation.

One ounce, by measure, should dissolve at least thirteen grains of white marble.

It should not form a precipitate on the addition of a solution of acetate of barytes, or of water saturated with sulphuretted hydrogen. The former circumstance shews, that it is adulterated with sulphuric acid; and the latter indicates a metal.

The metallic impregnation is best rendered obvious by sulphuretted hydrogen, in the manner stated. The distilled vinegar of commerce usually contains tin, and not lead, as has been asserted.

Adulteration of Cream.

Cream is often adulterated with rice powder or arrow root. The former is frequently employed for that purpose by pastry cooks, in fabricating creams and custards, for tarts, and other kinds of pastry. The latter is often used in the London dairies. Arrow-root is preferable to rice powder; for, when converted with milk into a thick mucilage by a gentle ebullition, it imparts to cream, previously diluted with milk, a consistence and apparent richness, by no means unpalatable, without materially impairing the taste of the cream.

The arrow-root powder is mixed up with a small quantity of cold skimmed milk into a perfect, smooth, uniform mixture; more milk is then added, and the whole boiled for a few minutes, to effect the solution of the arrow-root: this compound, when perfectly cold, is mixed up with the cream. From 220 to 260 grains, (or three large tea-spoonfuls) of arrow root are added to one pint of milk; and one part of this solution is mixed with three of cream. It is scarcely necessary to state that this sophistication is innocuous.

The fraud may be detected by adding to a tea-spoonful of the sophisticated cream a few drops of a solution of iodine in spirit of wine, which instantly produces with it a dark blue colour. Genuine cream acquires, by the addition of this test, a faint yellow tinge.

Poisonous Confectionery.

In the preparation of sugar plums, comfits, and other kinds of confectionery, especially those sweetmeats of inferior quality, frequently exposed to sale in the open streets, for the allurement of children, the grossest abuses are committed. The white comfits, called sugar pease, are chiefly composed of a mixture of sugar, starch, and Cornish clay (a species of very white pipe-clay;) and the red sugar drops are usually coloured with the inferior kind of vermilion. The pigment is generally adulterated with red lead. Other kinds of sweetmeats are sometimes rendered poisonous by being coloured with preparations of copper. The following account of Mr. Miles may be advanced in proof of this statement.

"Some time ago, while residing in the house of a confectioner, I noticed the colouring of the green fancy sweetmeats being done by dissolving sap-green in brandy. Now sap-green itself, as prepared from the juice of the buckthorn berries, is no doubt a harmless substance; but the manufacturers of this colour have for many years past produced various tints, some extremely bright, which there can be no doubt are effected by adding preparations of copper.

"The sweetmeats which accompany these lines you will find exhibit vestiges of being contaminated with copper.—The practice of colouring these articles of confectionery should, therefore, be banished: the proprietors of which are not aware of the deleterious quality of the substances employed by them."

The foreign conserves, such as small green limes, citrons, hop-tops, plums, angelica roots, &c. imported into this country, and usually sold in round chip boxes, are frequently impregnated with copper.

The adulteration of confitures by means of clay, may be detected by simply dissolving the comfits in a large quantity of boiling water. The clay, after suffering the mixture to stand undisturbed for a few days, will fall to the bottom of the vessel; and on decanting the clear fluid, and suffering the sediment to become dry gradually, it may be obtained in a separate state. If the adulteration has been effected by means of clay, the obtained precipitate, on exposure to a red heat in the bowl of a common tobacco-pipe, acquires a brick hardness.

The presence of copper may be detected by pouring over the comfits liquid ammonia, which speedily acquires a blue colour, if this metal be present. The presence of lead is rendered obvious by water impregnated with sulphuretted hydrogen, acidulated with muriatic acid which assumes a dark brown or black colour, if lead be present.

Poisonous Catsup.

This article is very often subjected to one of the most reprehensible modes of adulteration ever devised. Quantities are daily to be met with, which, on a chemical examination, are found to abound with copper. Indeed, this condiment is often nothing else than the residue left behind after the process employed for obtaining distilled vinegar, subsequently diluted with a decoction of the outer green husk of the walnut, and seasoned with all-spice, Cayenne pepper, pimento, onions, and common salt.

The quantity of copper which we have, more than once, detected in this sauce, used for seasoning, and which, on account of its cheapness, is much resorted to by people in the lower walks of life, has exceeded the proportion of lead to be met with in other articles employed in domestic economy.

The following account of Mr. Lewis on this subject, will be sufficient to cause the public to be on their guard.

"Being in the habit of frequently purchasing large quantities of pickles and other culinary sauces, for the use of my establishment, and also for foreign trade, it fell lately to my lot to purchase from a manufacturer of those commodities a quantity of walnut catsup, apparently of an excellent quality; but, to my great surprise, I had reason to believe that the article might be contaminated with some deleterious substance, from circumstances which happened in my business as a tavern keeper, but which are unnecessary to be detailed here; and it was this that induced me to make inquiry concerning the compounding of the suspected articles.

"The catsup being prepared by boiling in a copper, as is usually practised, the outer green shell of walnuts, after having been suffered to turn black on exposure to air, in combination with common salt, with a portion of pimento and pepper-dust, in common vinegar, strengthened with some vinegar extract, left behind as residue in the still of vinegar manufacturers; I therefore suspected that the catsup might be impregnated with some copper. To convince myself of this opinion. I boiled down to dryness a quart of it in a stone pipkin, which yielded to me a dark brown mass. I put this mass into a crucible, and kept it in a coal fire, red-hot, till it became reduced to a porous black charcoal; on urging the heat with a pair of bellows, and stirring the mass in the crucible with the stem of a tobacco-pipe, it became, after two hours' exposure to an intense heat, converted into a greyish-white ash; but no metal could be discriminated amongst it. I now poured upon it some aqua fortis, which dissolved nearly the whole of it, with an effervescence; and produced, after having been suffered to stand, to let the insoluble portion subside, a bright grass-green solution, of a strong metallic taste; after immersing into this solution the blade of a knife, it became instantly covered with a bright coat of copper.

"The walnut catsup was therefore evidently strongly impregnated with copper. On informing the manufacturer of this fact, he assured me that the same method of preparing the liquor was generally pursued, and that he had manufactured the article in a like manner for upwards of twenty years.

"Such is the statement I wish to communicate; and if you will allow it a place in your Literary Chronicle, it may perhaps tend to put the unwary on their guard against the practice of preparing this sauce by boiling it in a copper, which certainly may contaminate the liquor, and render it poisonous."

Poisonous Custard.

The leaves of the cherry laurel, *prunus lauro-cerasus,* a poisonous plant, have a nutty flavour, resembling that of the kernels of peach-stones, or of bitter almonds, which to most palates is grateful. These leaves have for many years been in use among cooks, to communicate an almond or kernel-like flavour to custards, puddings, creams, *blanc-mange,* and other delicacies of the table.

It has been asserted, that the laurel poison in custards and other articles of cookery is, on account of its being used in very small quantities, quite harmless. To refute this assertion, numerous instances might be cited; and, among them, a recent one, in which four children suffered most severely from partaking of custard flavoured with the leaves of this poisonous plant.

"Several children at a boarding-school, in the vicinity of Richmond, having partaken of some custard flavoured with the leaves of the cherry laurel, as is frequently practised by cooks, four of the poor innocents were taken severely ill in consequence. Two of them, a girl six years of age, and a boy of five years old, fell into a profound sleep, out of which they could not be roused.

"Notwithstanding the various medical exertions used, the boy remained in a stupor ten hours; and the girl nine hours; the other two, one of which was six years old, a girl, and a girl of seven years, complained of severe pains in the epigastric region. They all recovered, after three days' illness. I am anxious to communicate to you this fact, being convinced that your publication is read at all the scholastic establishments in this part of the country. I hope you will allow these lines a corner in your Literary Chronicle, where they may contribute to put the unwary on their guard, against the deleterious effects of flavouring culinary dishes with that baneful herb, the Cherry Laurel.

"I am, with respect, your's, Sir,

"Thomas Lidiard."

What person of sense or prudence, then, would trust to the discretion of an ignorant cook, in mixing so dangerous an ingredient in his puddings and creams? Who but a maniac would choose to season his victuals with poison?

The water distilled from cherry laurel leaves is frequently mixed with brandy and other spiritous liquors, to impart to them the flavour of the cordial called *noyeau,*

This fluid, though long in frequent use as a flavouring substance, was not known to be poisonous until the year 1728; when the sudden death of two women, in Dublin, after drinking some of the common distilled cherry laurel water, demonstrated its deleterious nature.

Poisonous Anchovy Sauce.

Several samples which we have examined of this fish sauce have been found contaminated with lead.

The mode of preparation of this fish sauce, consists in rubbing down the broken anchovy in a mortar: and this triturated mass, being of a dark brown colour, receives, without much risk of detection, a certain quantity of Venetian red, added for the purpose of colouring it, which, if genuine, is an innocent colouring substance; but instances have occurred of this pigment having been adulterated with orange lead, which is nothing else than a better kind of minium, or red oxide of lead. The fraud may be detected, as stated.

The conscientious oilmen, less anxious with respect to colour, substitute for this poison the more harmless pigment, called Armenian bole.

"Anchovies, 2 lbs. to 4 lbs. and a half; pulp through a fine hair sieve; boil the bones with common salt, 7 oz. in water 6 lbs.; strain; add flour 7 oz. and the pulp of the fish; boil; pass the whole through the sieve; colour with Venetian red to your fancy. It should produce one gallon."

Adulteration of Lozenges.

Lozenges, particularly those into the composition of which substances enter that are not soluble in water, as ginger, cremor tartar, magnesia, &c., are often sophisticated. The adulterating ingredient is usually pipe-clay, of which a liberal portion is substituted for sugar. The following detection of this fraud was lately made by Dr. T. Lloyd.

"Some ginger lozenges having lately fallen into my hands, I was not a little surprised to observe, accidentally, that when thrown into a coal fire, they suffered but little change. If one of the lozenges was laid on a shovel, previously made red-hot, it speedily took fire; but, instead of burning with a blaze and becoming converted into a charcoal, it took fire, and burnt with a feeble flame for scarcely half a minute, and there remained behind a stony hard substance, retaining the form of the lozenge. This unexpected result led me to examine these lozenges, which were bought at a respectable chemist's shop in the city; and I soon became convinced, that, in the preparation of them, a considerable quantity of common pipe-clay had been substituted for sugar. On making a complaint about this fraud at the shop where the article was sold, I was informed that there were two kinds of ginger lozenges kept for sale, the one at three-pence the ounce, and the other at six-pence per ounce; and that the article furnished to me by mistake was the cheaper commodity: the latter were distinguished by the epithet *verum*, they being composed of sugar and ginger only; but the former were manufactured partly of white Cornish clay, with a portion of sugar only, with ginger and Guinea pepper. I was likewise informed, that of Tolu lozenges, peppermint lozenges and ginger pearls, and several other sorts of lozenges, two kinds were kept; that the *reduced* articles, as they were called, were manufactured for those very clever persons in their own conceit, who are fond of haggling, and insist on buying better bargains than other people, shutting their eyes to the defects of an article, so that they can enjoy the delight of getting it cheap; and, secondly for those persons, who being but bad paymasters, yet, as the manufacturer, for his own credit's sake, cannot charge more than the usual price of the articles, he thinks himself therefore authorised to adulterate it in value, to make up for the risk he runs, and the long credit he must give."

The comfits called ginger pearls, are frequently adulterated with clay. These frauds may be detected in the manner stated.

Poisonous Olive Oil.

This commodity is sometimes contaminated with lead, because the fruit which yields the oil is submitted to the action of the press between leaden plates; and it is, moreover, a practice (particularly in Spain) to suffer the oil to become clear in leaden cisterns, before it is brought to market for sale. The French and Italian olive oil is usually free from this impregnation.

Olive oil is sometimes mixed with oil of poppy seeds: but, by exposing the mixture to the freezing temperature, the olive oil freezes, while that of the poppy seeds remains fluid; and as oils which freeze with most difficulty are most apt to become rancid, olive oil is deteriorated by the mixture of poppy oil.

Good olive oil should have a pale yellow colour, somewhat inclining to green; a bland taste, without smell; and should congeal at 38° Fahrenheit. In this country, it is frequently met with rancid.

The presence of lead is detected by shaking, in a stopped vial, one part of the suspected oil, with two or three parts of water impregnated with sulphuretted hydrogen. This agent will render the oil of a dark brown or black colour, if any metal, deleterious to health, be present. The practice of keeping this oil in pewter or leaden cisterns, as is often the case, is objectionable; because the oil acts upon the metal. The dealers in this commodity assert, that it prevents the oil from becoming rancid: and hence some retailers often suffer a pewter measure to remain immersed in the oil.

Adulteration of Mustard.

Genuine mustard, either in powder, or in the state of a paste ready for use, is perhaps rarely to be met with in the shops. The article sold under the name of *genuine Durham mustard*, is usually a mixture of mustard and common wheaten flour, with a portion of Cayenne pepper, and a large quantity of bay salt, made with water into a paste, ready for use. Some manufacturers adulterate their mustard with radish-seed and pease flour.

It has often been stated, that a fine yellow colour is given to mustard by means of turmeric. We doubt the truth of this assertion. The presence of the minutest quantity of turmeric may instantly be detected, by adding to the mustard a few drops of a solution of potash, or any other alkali, which changes the bright yellow colour, to a brown or deep orange tint.

Two ounces and a half of Cayenne pepper, 1-1/2 lbs. of bay salt, 8 lbs. of mustard flour, and 1-1/2 lbs. of wheaten flour, made into a stiff paste, with the requisite quantity of water, in which the bay-salt is previously dissolved, forms the so-called *genuine Durham mustard*, sold in pots. The salt and Cayenne pepper contribute materially to the keeping of ready-made mustard.

There is therefore nothing deleterious in the usual practice of adulterating this commodity of the table. The fraud only tends to deteriorate the quality and flavour of the genuine article itself.

Adulteration of Lemon Acid.

It is well known to every one, that the expressed juice of lemons is extremely apt to spoil, on account of the sugar, mucilage, and extractive matter which it contains; and hence various means have been practised, with the intention of rendering it less perishable, and less bulky. The juice has been evaporated to the consistence of rob; but this always gives an unpleasant empyreumatic taste, and does not separate the foreign matters, so that it is still apt to spoil when agitated on board of ship in tropical climates. It has been exposed to frost, and part of the water removed under the form of ice; but this is liable to all the former objections; and, besides, where lemons are produced in sufficient quantity, there is not a sufficient degree of cold. The addition of a portion of spirit to the inspissated juice, separates the mucilage, but not the extractive matter and the sugar. By means, however, of separating the foreign matters associated with it, in the juice, by chemical processes unnecessary to be detailed here, citric acid is now manufactured, perfectly pure, and in a crystallised form, and is sold under the name of concrete lemon acid. In this state it is extremely convenient, both for domestic and medicinal purposes. One drachm, when dissolved in one ounce of water, is equal in strength to a like bulk of fresh lemon juice. To communicate the lemon flavour, it is only necessary to rub a lump of sugar on the rind of a lemon to become impregnated with a portion of the essential oil of the fruit, and to add the sugar to the lemonade, negus, punch, shrub, jellies or culinary sauces, prepared with the pure citric acid.

Fraudulent dealers often substitute the cheaper tartareous acid, for citric acid. The negus and lemonade made by the pastry-cooks, and the liquor called punch, sold at taverns in this metropolis, is usually made with tartareous acid.

To discriminate citric acid from tartareous acid, it is only necessary to add a concentrated solution of the suspected acid, to a concentrated solution of muriate of potash, taking care that the solution of the acid is in excess. If a precipitate ensues, the fraud is obvious, because citric acid does not produce a precipitate with a solution of muriate or potash.

Or, by adding to a saturated solution of tartrate of potash, a saturated solution of the suspected acid, in excess, which produces with it an almost insoluble precipitate in minute granular crystals. Pure citric acid produces no such effect when added in excess to tartrate of potash.

Poisonous Mushrooms.

Mushrooms have been long used in sauces and other culinary preparations; yet there are numerous instances on record of the deleterious effects of some species of these *fungi*, almost all of which are fraught with poison. Pliny already exclaims against the luxury of his countrymen in this article, and wonders what extraordinary pleasure there can be in eating such dangerous food.

But if the palate must be indulged with these treacherous luxuries, or, as Seneca calls them, "voluptuous poison," it is highly necessary that the mild eatable mushrooms, should be gathered by persons skilful enough to distinguish the good from the false, or poisonous, which is not always the case; nor are the characters which distinguish them strongly marked.

The following statement is published by Mr. Glen, surgeon, of Knightsbridge:

"A poor man, residing in Knightsbridge, took a walk in Hyde Park, with the intention of gathering some mushrooms. He collected a considerable number, and, after stewing them, began to eat them. He had finished the whole, with the exception of about six or eight, when, about eight or ten minutes from the commencement of his meal, he was suddenly seized with a dimness, or mist before his eyes, a giddiness of the head, with a general trembling and sudden loss of power;—so much so, that he nearly fell off the chair; to this succeeded loss of recollection: he forgot where he was, and all the circumstances of his case. This deprivation soon went off, and he so far rallied as to be able, though with difficulty, to get up, with the intention of going to Mr. Glen for assistance—a distance of about five hundred yards: he had not proceeded more than half way, when his memory again failed him; he lost his road, although previously well acquainted with it. He was met by a friend, who with difficulty learned his state, and conducted him to Mr. Glen's house. His countenance betrayed great anxiety: he reeled about, like a drunken man, and was greatly inclined to sleep; his pulse was low and feeble. Mr. Glen immediately gave him an emetic draught. The poison had so diminished the sensibility of the stomach, that vomiting did not take place for near twenty minutes, although another draught had been exhibited. During this interval his drowsiness increased to such a degree, that he was only kept awake by obliging him to walk round the room with assistance; he also, at this time, complained of distressing pains in the calves of his legs.— Full vomiting was at length produced. After the operation of the emetic, he expressed himself generally better, but still continued drowsy. In the evening Mr. Glen found him doing well."

The following case is recorded in the Medical Transactions, vol. ii.

"A middle-aged man having gathered what he called champignons, they were stewed, and eaten by himself and his wife; their child also, about four years old, ate a little of them, and the sippets of bread which were put into the liquor. Within five minutes after eating them, the man began to stare in an unusual manner, and was unable to shut his eyes. All objects appeared to him coloured with a variety of colours. He felt a palpitation in what he called his stomach; and was so giddy, that he could hardly stand. He seemed to himself swelled all over his body. He hardly knew what he did or said; and sometimes was unable to speak at all. These symptoms continued in a greater or less degree for twenty-four hours; after which, he felt little or no disorder. Soon after he perceived himself ill, one scruple of white vitriol was given him, and repeated two or three times, with which he vomited plentifully.

"The woman, aged thirty-nine, felt all the same symptoms, but in a higher degree. She totally lost her voice and her senses, and was either stupid, or so furious that it was necessary she should be held. The white vitriol was offered to her, of which she was capable of taking but very little; however, after four or five hours, she was much recovered: but she continued many days far from being well, and from enjoying her former health and strength. She frequently fainted for the first week after; and there was, during a month longer, an uneasy sense of heat and weight in her breast, stomach, and bowels, with great flatulence. Her head was, at first waking, much confused; and she often experienced palpitations, tremblings, and other hysteric affections, to all which she had ever before been a stranger.

"The child had some convulsive agitations of his arms, but was otherwise little affected. He was capable of taking half a scruple of ipecacuanha, with which he vomited, and was soon perfectly recovered."

MUSHROOM CATSUP.

The edible mushroom is the basis of the sauce called mushroom catsup; a great proportion of which is prepared by gardeners who grow the fungi. The mushrooms employed for preparing this sauce are generally those which are in a putrefactive state, and not having found a ready sale in the market; for no vegetable substance is liable to so rapid a spontaneous decomposition as mushrooms. In a few days after the fungus has been removed from the dung-bed on which it grows, it becomes the habitation of myriads of insects; and, if even the saleable mushroom be attentively examined, it will frequently be found to swarm with life.

Poisonous Soda Water.

The beverage called soda water is frequently contaminated both with copper and lead; these metals being largely employed in the construction of the apparatus for preparing the carbonated water, and the great excess of carbonic acid which the water contains, particularly enables it to act strongly on the metallic substances of the apparatus; a truth, of which the reader will find no difficulty in convincing himself, by suffering a stream of sulphuretted hydrogen gas to pass through the water.

Food poisoned by Copper Vessels.

Many kinds of viands are frequently impregnated with copper, in consequence of the employment of cooking utensils made of that metal. By the use of such vessels in dressing food, we are daily liable to be poisoned; as almost all acid vegetables, as well as sebaceous or pinguid substances, employed in culinary preparations, act upon copper, and dissolve a portion of it; and too many examples are met with of fatal consequences having ensued from eating food which had been dressed in copper vessels not well cleaned from the oxide of copper which they had contracted by being exposed to the action of air and moisture.

The inexcusable negligence of persons who make use of copper vessels has been productive of mortality, so much more terrible, as they have exerted their action on a great number of persons at once. The annals of medicine furnish too many examples in support of this assertion, to render it necessary to insist more upon it here.

Mr. Thiery, who wrote a thesis on the noxious quality of copper, observes, that "our food receives its quantity of poison in the kitchen by the use of copper pans and dishes. The brewer mingles poison in our beer, by boiling it in copper vessels. The sugar-baker employs copper pans; the pastry-cook bakes our tarts in copper moulds; the confectioner uses copper vessels: the oilman boils his pickles in copper or brass vessels, and verdigris is plentifully formed by the action of the vinegar upon the metal.

"Though, after all, a single dose be not mortal, yet a quantity of poison, however small, when taken at every meal, must produce more fatal effects than are generally apprehended; and different constitutions are differently affected by minute quantities of substances that act powerfully on the system."

The author of a tract, entitled, "Serious Reflections on the Dangers attending the Use of Copper Vessels," asserts that a numerous and frightful train of diseases is occasioned by the poisonous effects of pernicious matter received into the stomach insensibly with our victuals.

Dr. Johnston gives an account of the melancholy catastrophe of three men being poisoned, after excruciating sufferings, in consequence of eating food cooked in an unclean copper vessel, on board the Cyclops frigate; and, besides these, thirty-three men became ill from the same cause.

The following case is related by Sir George Baker, M. D.

"Some cyder, which had been made in a gentleman's family, being thought too sour, was boiled with honey in a brewing vessel, the rim of which was capped with lead. All who drank this liquor were seized with a bowel colic, more or less violently. One of the servants died very soon in convulsions; several others were cruelly tortured a long time. The master of the family, in particular, notwithstanding all the assistance which art could give him, never recovered his health; but died miserably, after having almost three years languished under a most tedious and incurable malady."

Too much care and attention cannot be taken in preserving all culinary utensils of copper, in a state unexceptionably fit for their destined purpose. They should be frequently tinned, and kept thoroughly clean, nor should any food ever be suffered to remain in them for a longer time than is absolutely necessary to their preparation for the table. But the sure preventive of its pernicious effect, is, to banish copper utensils from the kitchen altogether.

The following wholesome advice on this subject is given to cooks by the author of an excellent cookery book.

"Stew-pans and soup-kettles should be examined every time they are used; these, and their covers, must be kept perfectly clean and well tinned, not only on the inside, but about a couple of inches on the outside; so much mischief arises from their getting out of repair; and, if not kept nicely tinned, all your work will be in vain; the broths and soups will look green and dirty, and taste bitter and poisonous, and will be spoiled both for the eye and palate, and your credit will be lost; and as the health, and even the life, of the family depends upon this; the cook may be sure her employer had rather pay the tin-man's bill than the doctor's."

The senate of Sweden, in the year 1753, prohibited copper vessels, and ordered that none but such as were made of iron should be used in their fleet and armies.

Food Poisoned by Leaden Vessels.

Various kinds of food used in domestic economy, are liable to become impregnated with lead.

The glazing of the common cream-coloured earthen ware, which is composed of an oxide of lead, readily yields to the action of vinegar and saline compounds; and therefore jars and pots of this kind of stone ware, are wholly unfit to contain jellies of fruits, marmalade, and similar conserves. Pickles should in no case be deposited in cream-coloured glazed earthenware.

The custom which still prevails in some parts of this country of keeping milk in leaden vessels for the use of the dairy, is very improper.

"In Lancashire the dairies are furnished with milk-pans made of lead: and when Mr. Parks expostulated with some individuals on the danger of this practice, he was told that *leaden* milk-pans throw up the cream much better than vessels of any other kind.

"In some parts of the north of England it is customary for the inn-keepers to prepare mint-salad by bruising and grinding the vegetable in a large wooden bowl with a *ball of lead* of twelve or fourteen pounds weight. In this operation the mint is cut, and portions of the lead are ground off at every revolution of the ponderous instrument. In the same county, it is a common practice to have brewing-coppers constructed with the bottom of copper and the whole sides of lead."

The baking of fruit tarts in cream-coloured earthenware, and the salting and preserving of meat in leaden pans, are no less objectionable. All kinds of food which contain free vegetable acids, or saline preparations, attack utensils covered with a glaze, in the composition of which lead enters as a component part. The leaden beds of presses for squeezing the fruit in cyder countries, have produced incalculable mischief. These consequences never follow, when the lead is combined with tin; because this metal, being more eager for oxidation, prevents the solution of the lead.

When we consider the various unsuspected means by which the poisons of lead and copper gain admittance into the human body, a very common but dangerous instance presents itself: namely, the practice of painting toys, made for the amusement of children, with poisonous substances, viz. red lead, verdigris, &c. Children are apt to put every thing, especially what gives them pleasure, into their mouths; the painting of toys with colouring substances that are poisonous, ought therefore to be abolished; a practice which lies the more open to censure, as it is of no real utility.